Ismael Noronha de Oliveira
Dorgival Júnior
Patrícia Araújo

Modeling and Development of an Autopilot Prototype

Ismael Noronha de Oliveira
Dorgival Júnior
Patrícia Araújo

Modeling and Development of an Autopilot Prototype

Control and Dynamics

ScienciaScripts

Imprint

Any brand names and product names mentioned in this book are subject to trademark, brand or patent protection and are trademarks or registered trademarks of their respective holders. The use of brand names, product names, common names, trade names, product descriptions etc. even without a particular marking in this work is in no way to be construed to mean that such names may be regarded as unrestricted in respect of trademark and brand protection legislation and could thus be used by anyone.

Cover image: www.ingimage.com

This book is a translation from the original published under ISBN 978-613-9-61957-3.

Publisher:
Sciencia Scripts
is a trademark of
Dodo Books Indian Ocean Ltd. and OmniScriptum S.R.L publishing group

120 High Road, East Finchley, London, N2 9ED, United Kingdom
Str. Armeneasca 28/1, office 1, Chisinau MD-2012, Republic of Moldova, Europe
Printed at: see last page
ISBN: 978-620-7-70783-6

SUMMARY

Dedication,

I dedicate this work to God, for He is worthy of all honor and glory, He has taken care of me.

To my mother for all her love and care for me.

To my brothers Gercivânia, Daniel, Samuel and Mizael.

To my maternal grandparents Antônia Noronha and Francisco das Chagas.

And to all my family.

"Lord, I want to thank you with all my heart and speak of all your wonders".

(Psalms 9:1)

ACKNOWLEDGMENTS

I want to thank God first of all, because his mercies are renewed every morning in my life.

My mother, who always fights to give me and my siblings the best she can, never spared any effort to help us, was, is and will always be a warrior every day of her life.

To my sister and brothers Gercivânia, Daniel, Samuel and Mizael, who have always fought and are fighting to achieve their goals.

To my maternal grandparents, Francisco das Chagas and Antônia Noronha, for always passing on love, discipline and respect to their descendants.

To my family for always putting their trust in my goals, especially: Geralda, Daluz, Deide, Regina, Lucia, Lenildo, Luiz, Jailson, Jean, Rayane, Ana Claudia, Jhonnata Gabriel and Marcelo.

To my teachers who made a difference in my life during all my years of study, in particular: Otàvio Paulino for being my source of inspiration when I started the course, Jorge Pinto for helping me through one of the most difficult stages of my degree, Professor André Rocha for passing on his knowledge with excellence and Professor Fernando Grings for teaching me Mathematics with mastery.

To my teacher and advisor Dorgival Albertino for his patience, guidance, teachings, inspiration and confidence in my work.

To my friend Ricardo César, for his companionship, advice, prayers, affection and for becoming the father I never had.

To my friend Patricia Rodrigues, for her patience, her teaching, for passing on her knowledge, for dedicating her time to helping me carry out this work.

To my fellow students, especially: Amim Alleff, Valéria, Bruno José, for the late nights we spent studying together,

To my partners in Mossoró, Marcelo Virginio, Ronier Silva, Filipe Oliveira, Pedro Cândido, you were fundamental in getting me this far.

SUMMARY

This work deals with the manufacture of an electronic throttle designed in the three-dimensional modeling *software* Autodesk Inventor 2015 and mechanically manufactured from AISI 304 steel plates and billets. The aim was to design the pedal in a similar way to car pedals. The electronic project was developed using Proteus electronic circuit *software,* serving as a reference for assembling the LCD *display*, *encoder,* transistor, capacitor, diode, potentiometer, battery and all the wiring. Each component in this project is connected to the Arduino® UNO board, where it interprets the status of each *setpoint* and drives the power circuit and motor. The *encoder* sensor attached to the motor shaft gives a reading of the rotation per minute (RPM). The electronic throttle is programmed in C/C++, which has been tested and inserted into the Arduino® board to manage the speed controller's activities by relating the pedal position to the tachometer's rotation. The implemented code controls the speed electronically by pressing buttons to increase or decrease the motor's rotation speed. The linear proportionality relationship between the tachometer's rotation speed and the voltage was evident in the results, as was the linear relationship between the voltage and the angular measurement of the pedal's displacement. The feasibility of the project proved to be efficient given the low cost of the materials used.

KEYWORDS: Mechanics. Cars. Accelerator. Electronic. Arduino.

ACRONYMS AND SYMBOLS

SIGLA	MEANING
IDE	Integrated Development Environment
CPU	*Central* Processing Unit
USB	*Universal Serial Bus* (Universal Port)
LED	*Light* Emitting Diode
SPI	Serial Peripheral Interface
KB	Kilobyte
RFID	Radio-Frequency IDentification
ANPR	Automatic License Plate Recognition
RF	Radio Frequency
MHz	Mega Hertz
LCD	Liquid Crystal Display
EEG	Electroencephalogram
FED	*Experimental* Feeding Device
PNP	Positive-Negative-Positive
NPN	Negative-Positive-Negative
TBJ	*Bipolar Junction Transistor* (BJT)
PWM	*Pulse Width Modulation*
CC	Direct Current

SYMBOLOGY	DESCRIPTION
β or hfe	Current gain
I_E	Emitter current
I_C	Collector current
I_B	Base chain
V_{BE}	Base Emitter Voltage
V	Volts
R_B	Base resistance
R_C	Collector resistance
Vdc	Time the wave stays at a high level
Vpulse	PWM signal pulse voltage
Vdc	Average voltage of a waveform

1   INTRODUCTION

It is notable that in recent decades there has been an expansion in the production of knowledge, with countless political and economic changes in societies around the world. This has been due to the introduction of various technological innovations that have enabled the universalization of information, making it possible for the population to have almost real-time information on what is happening anywhere on the planet. With the development of knowledge, all areas of information have grown, and technology has been no different. With the world becoming increasingly automated, the area of programming has played a significant role in the development of new technologies (AGUIA; PASSOS, 2014).

Since programming and electronics are areas of research for many students around the world and research and prototyping materials are not cheap, Massimo Banzi and other researchers at the Ivrea Institute of Interaction *Design in* Italy in 2005 developed a board consisting of an 8-bit Atemga microcontroller with complementary components.with open source code and a programming language based on C/C++; helping all students to develop prototyping systems in electronics (EVANS; NOBLE; HOCHENBAUM, 2013, p.25).

Arduino® is used to solve the most diverse problems, one of which was to develop a security system to control access for people and cars in condominiums. This is how Almeida and Luz (2014) developed a project to improve the security system of condominiums, with the aim of developing a unified access control system that would mainly deal with the security of residents using technologies such as Radio Frequency Identification (RFID) and Automatic Vehicle Plate Recognition (ANPR). The use of technologies such as Arduino® in the area of security automation provides easy, fast and inexpensive control in a residential environment (ALMEIDA; LUZ, 2014).

Arduino® is also used to develop systems to combat corrosion. To do this, it is necessary to carry out an investigation using different physical-chemical techniques, which provide information on the composition, microstructure and morphology of corrosion products. An innovative, portable, low-cost and easy-to-use system was developed by S. Grassinia et al. (2016), based on the Arduino® board and specifically designed measuring probes. The tool developed is an important tool for conservators and art historians to use Arduino® to find a way of monitoring the stability of their artifacts, as it is difficult for them to find tools to help with their conservation (S. Grassini et al., 2016).

The major car manufacturers invest in research to develop new technologies for their cars, which are directly linked to the driver, but what automotive companies are looking for today are modernizations that make cars increasingly independent, providing more comfort and economy for drivers (FREITA et al., 2009).

For future work, it is proposed to develop a braking system and implement it in the system already designed. Another proposal would be to design a closed-loop system, which would allow greater control of the system.

This paper aims to show the development of an electronic throttle system in a didactic way.

1.1 OBJECTIVES

The aim of this work is to develop a prototype electronic accelerator for a car, using AISI 304 steel and an Arduino® UNO board.

The specific objectives were:

a)   Designing a pedal that simulates a car;

b)   Develop a prototype electronic accelerator using AISI 304 steel;

c)   Develop an automatic pilot system in the project;

d)   Implement an anti-corrosion system in the pedal.

1.2 WORK ORGANIZATION

The work is organized as follows:

Chapter 2 contains a bibliographical review of mechanics applied to engineering, technical drawing, corrosion, the Arduino® system, historical context, examples of various applications of Arduino® boards; the speed control system; the laws of physics that explain how an electrical circuit works and the method for calculating armature resistance.

Chapter 3 presents the description, the materials used and the control flowchart for the autopilot developed.

Chapter 4 contains the experimental results of the sensors and actuators. A graph was drawn up showing the relationship between the supply voltage and the speed in RPM of the motor, and graphs were also generated showing the relationship between the angle and the output voltage. The total cost of the prototype and its feasibility were also analyzed.

Chapter 5 then describes the conclusion of this work and presents proposals for future work.

2   LITERATURE REVIEW

2.1   MECHANICS

Mechanics is the part of science that studies the state of motion or rest of bodies that are subject to the action of forces. When it comes to engineering applications, the design area of mechanics is divided into two: statics and dynamics. Static relates to the equilibrium of a body that is stationary or moving at a constant speed. Dynamics refers to the accelerated movement of a body, and is divided into two parts: kinematics, which is linked to the geometry of movement, and kinetics, which analyzes the forces that cause displacement (HIBBELER, 2011, p.1).

The principles of dynamics were developed when it became possible to measure time accurately. Galileo Galilei (1564-1642) was one of the pioneers among the most important contributors to this field of study. He carried out experiments using pendulums and falling bodies. The most important contributions to dynamics were made by Isaac Newton (1642-1727), who is best known for formulating his three fundamental laws of motion and the law of universal gravitational attraction. After Isaac Newton's discoveries, other important techniques for application in engineering studies were developed by D'Alembert, Lagrange, Euler and others (HIBBELER, 2011, p. 1).

In kinematics, the movement of a rigid body at any time t can be classified as the superposition of a translational movement with a rotational movement. The translational displacement can have the momentary velocity of any point on the body, and the rotational velocity, ω, has its axis of rotation passing through the given point. Thus, from the dynamics of a particle, we can describe the movement of translation. The movement of the center of mass of any set of particles is relative to the total external force, which is described in equation 1 (SHAMES, 2003, p. 337).

$$F = M.A \tag{1}$$

Where:

A is the acceleration of the particle,

M is the total mass of the particle system.

The position of a particle is defined by the position vector r, in which this vector extends from its origin to its end, if the body rotates $d\theta$, then equation 2 describes the displacement of the particle.

$$s = rd\theta \tag{2}$$

When equation 2 is derived, the particle's velocity, described in equation 3, is found.

$$v = \omega r \tag{3}$$

The particle's acceleration can be expressed in two components, the normal acceleration and the tangential acceleration, which are described in equations 4 and 5 respectively.

$$a_n = \omega^2 r \qquad\qquad (4)$$

$$a_t = \alpha r \qquad\qquad (5)$$

2.2 TECHNICAL DRAWING

The main objective of technical drawing is the precise representation, on a plane, of the shapes of the material world, and therefore three-dimensional, in such a way that they can be spatially reassembled. According to Pertence, Santos and Jardim (2001), approximately one fifth of the errors found in the work of technical drawing students are due to a lack of spatial intelligence skills.

One of the requirements for a good performance when drawing up projects is a good geometric interpretation and spatial intelligence, so that you can get a sense of the representation of the parts to be designed. In this way, the designer can easily represent a three-dimensional part on a two-dimensional plane and vice versa. Another key requirement is the interaction between the designer and the project execution team, as both parties need to have the same technical language in order to facilitate project execution (PERTENCE; SANTOS; JARDIM, 2001).

Technical drawing became a scientific discipline from the 18th century onwards, and its aim was to serve the industry of society at that time. To this end, geometric drawing was combined with projective and descriptive geometry to create a universal visual language, in which the designer provides the information that the project execution team needs to manufacture the object. For this to happen in harmony, a complex system of conventions is followed which must be respected by the designer who registers it, as well as by the manufacturer (SOARES, 2007).

Technical drawing has become important for the graphic expression of projects, and all countries considered industrialized have regulated its creation through technical standards. There are technical standards that provide guidelines for designing a wide variety of objects. The international technical standards ISO 5457:1980 and ISO 216:1975 regulate the format and orientation of the paper on which the project will be drawn, based on the following sizes: A0, which has an area of 1.00 m2, A1, which has an area of 0.500 m2, A2, which has an area of 0.250 m^2 , A3, which has an area of 0.125 m^2 and A4, which has an area of 0.062 m^2 (FRANCESCONI, 2010).

The drawing sheet must contain a legend with the following information: the title of the exercise, the name of the designer, the name of the course and the institution or company in which he works, the symbol of the projection method, the scale of the drawing, the linear dimensional unit and the date the drawing was made. Depending on the part to be drawn, it is necessary to use several different lines. The ISO 128:1982 standard defines 10 types of lines. There are other standards that regulate how the project should be drawn up (FRANCESCONI, 2010).

Drawing can be done by hand, using the appropriate equipment: pencil, square, scalimeter, paper, compass, etc. You can also use a computer with appropriate drawing *software*. Technical drawing has become an indispensable tool for engineering, as complex projects with a combination of several parts have been better presented using drawing installation images. The growth was observed in the maritime sector, as the design and applications used in ship manufacturing also gained importance in the balance sheet and construction account (YAPICI; KOLDEMIRB, 2015).

In recent years, it has become easier to draw up complex projects, *software has* become more advanced and you can draw a part, machine or other object in great detail. All this can be done with better design quality and in less time. The most sophisticated *software already* has a series of materials included in its library, but it also allows you to import parts or objects from the internet so that the project has a better dynamic and a good presentation, which is the case with the three-dimensional modeling *software* Autodesk Inventor, in which it has in its library: screws, nuts, connections, among others. This makes it more attractive, easier and faster to make a technical drawing (YAPICI; KOLDEMIRB, 2015).

2.3   CORROSION

Since prehistoric times, when man was still learning to handle metals, he has been searching for a solution to a serious problem: corrosion. Despite all the development of science, even today corrosion causes major problems in the most varied activities, causing material damage and also endangering man's own safety. Corrosion can cause critical parts of equipment to suddenly break down, leading to accidents that can even result in loss of life. At home or at work, the corrosive process can be seen in doors, fires, information boards, fridges, cupboards and so on. There are three types of corrosion, which are classified as: electrochemical, chemical and electrolytic (FRAUCHES-SANTOS; ALBUQUERQUE; ECHEVARRIA, 2013).

Corrosion is defined as the total, partial, superficial or structural deterioration of a material, usually metallic, due to the physical, chemical or electrochemical action of the environment, whether or not combined with mechanical stress. In general, corrosion is a spontaneous process, and is often modifying metallic materials in such a way that their resistance and performance no longer meet the purposes for which they were intended. The deterioration caused by the physical-chemical interaction between the material and the environment in which it is found leads to harmful and undesirable changes borne by the material, such as degradation, chemical alterations or structural transformations, making the material unsuitable for use in some situations (FRAUCHES-SANTOS; ALBUQUERQUE; ECHEVARRIA, 2013).

Corrosion has been the cause of major problems in a wide variety of activities, especially in the chemical and oil industries, in rail, road, subway, air, sea and communications systems, among others. One way of combating corrosion in steel materials, in the raceways of the steel production process and in the steel industry is to add carbon to their composition (LUZ; RIBEIRO; PANDOLFELLI, 2008).

There are also other ways of combating corrosion, for example, creating an impermeable barrier between the steel and the corrosive environment, introducing a substance that inhibits the chemical action of corrosion, inducing a galvanic activity that acts against the galvanic activity of oxidation. Thin organic coatings are the most widely used for protection against oxidation because of their low cost in relation to the degree of protection obtained. To obtain a good paint system, it is necessary to clean the material properly, apply a base coat and then an outer coat (FRAUCHES-SANTOS; ALBUQUERQUE; ECHEVARRIA, 2013).

Because their structures are made of steel and because they are generally exposed to a corrosive environment, their structures are damaged by corrosion. Military vehicles work in various locations around the world and are linked to various types of structural failure influenced by corrosion. These means of transportation can be at sea, which is a very aggressive environment because salt is a corrosive agent, but they can also be in the desert, which is an environment full of corrosive particles. Due to this variation in location and the change in temperature and harsh working conditions, the components of these vehicles fail. After 18 years, the Canadian Army had 25% of its fleet with visible component failures caused by corrosion (SAEED; KHAN; NAZIR, 2016).

Researchers Saeed, Khan and Nazir (2016) carried out a study to evaluate the action of corrosion on a military car. A sample collected from the vehicle revealed surface corrosion mechanisms, subsurface propagation and contaminants. In simulated environmental tests, even in good environmental conditions, corrosion accumulation was noted. It was also found that optimum temperature and humidity levels need to be achieved in order to minimize the ageing process of these vehicles caused by corrosion. The surface of the car showed flaws in its paint system, thus facilitating the action of corrosion.

2.4 ARDUINO®

This section aims to show the various applications of the Arduino® board in different areas of study, thus demonstrating the capacity of a simple tool to work with and low-cost equipment.

The main components of the Arduino® are *hardware and software,* where the former consists of a prototyping board and the latter is an IDE (Integrated Development Environment), which is run on a computer to do the programming. Arduino® uses an 8-bit Atmega AVR microcontroller, which has components that facilitate programming and the inclusion of other circuits. The connectors on the board are laid out in such a way as to allow the CPU to be connected to other expandable modules, known as *shields* (EVANS; NOBLE; HOCHENBAUM, 2013, p.25).

Arduino® is used to create a wide variety of interactive objects, as it allows connection to sensors, motors, lights and other physical objects. The Arduino® board is an open source electronic prototyping platform. It can capture the state of the environment using sensors and interact with its surroundings by controlling actuators. Arduino® has its own programming language, based on C/C++ (EVANS; NOBLE; HOCHENBAUM, 2013, p.25).

Weather forecasting has always been an object of study for humankind. What is sought today are new practical ways of achieving a quick and effective result, and to this end, people are making use of the most diverse technologies to develop tools that help people's lives. Thus, the researchers Laskara et al. (2016) developed an autonomous cube satellite system using an Arduino® data processing unit and three sensors, namely a temperature sensor (DHT11), a pressure sensor (BMP085) and an accelerometer (ADXL-335). The data was recorded on a computer and transmitted using a cube-SAT, a monitoring device with a transmitter and receiver module. A 433 MHz RF module was used for this transmission data (LASKARA et al., 2016).k

In terms of biotechnology, the equipment used to extract venom from animals is expensive. Seeking a new alternative for extracting venom from arthropods, Besson et al. (2016) created a new, affordable device for extracting it. They used an Arduino® board and other equipment to manufacture the extractor. The system created had satisfactory results. The system designed by the researchers was distinguished by its ease of handling and low cost (BESSON et al., 2016).

The electroencephalogram (EEG) is a test that makes a graphic record of the spontaneous electrical currents developed in the brain, through electrodes applied to the scalp, on the brain surface. According to Kelmann and Bernardo (2012), "conventional EEG is based on visual examination of the tracing and therefore has a significant subjective component". Researchers Yu and Sim (2016) proposed an Arduino® method that classifies the electroencephalogram from color imagery data; using an *Emotiv EPOC* headset, an LED and a computer. It was observed that with the simple perception of a color by a patient, the machine was able to recognize the patient's thoughts. This study is relevant for helping people with paralysis and the elderly (YU; SIM, 2016).

Rodents have been used in laboratories in scientific experiments in order to contribute to the development of science and technology in the most diverse areas of knowledge, they have served as experimental models to observe behaviors, development, drug quality control, preclinical studies, basic sciences and experimental surgeries (MOURA, 2014). With this in mind, Nguyen et al. (2016) developed an Experimental Feeding Device (FED), this device releases a single pellet of food into a reservoir, which is monitored by infrared sensors, each time the rodent eats the pellet the time and day are recorded, then the device releases a new pellet. The EDF uses an Arduino® to control the system (NGUYEN et al., 2016).

Animals are often used in laboratories for scientific purposes, with the aim of developing technology to improve human life (MOURA, 2014). According to researchers Ponce et al. (2016) "Neuroscientific research on non-human primates usually requires the animals to sit in a chair. To do this, the monkeys are fitted with collars and trained to sit on chairs using a leash." They developed a method for training animals to sit in a primate chair for neck plate placement. They equipped the chair with an Arduino®, a water reward system and touch and proximity sensors. The results were satisfactory compared to previous methods, as the animals were not stressed or injured by this new method (PONCE et al., 2016).

Robot technology has advanced rapidly and today they can be found with the most diverse functions. In relation to this, Pandey et al. (2016) presented in their work an inference of the *Adaptive Neuro-Fuzzy Inference System (*ANFIS) controller, this controller is useful for mobile robots and obstacle avoidance in unknown static environments. They used an Arduino® board and some sensors: a rangefinder sensor, ultrasound and an infrared sensor. This inference obtained satisfactory results, as the robot did not collide with the obstacles (PANDEY et al., 2016).

Still on the robotics front, researchers Munawar et al. (2015) have developed an intelligent *Computer Retina* system, controlled with a wireless Zigbee base. The Zigbee board designates a set of specifications for wireless communication between electronic devices, with an emphasis on low operating power, low data transmission rates and low implementation costs. This system is an excellent solution for people with paralysis and other diseases that prevent them from moving their bodies. The system is attached to a pair of glasses that are placed on the user's head, and the movement of the patient's head serves as a mouse on a normal computer. This system stands out because it uses technology for a humanitarian function, providing help for people with disabilities, where hospital patients can interact with a computer (MUNAWAR et al., 2015).

2.5 SPEED CONTROL

The speed controller provides more autonomy for the car, more safety and comfort in adverse situations for drivers, so the driver of the vehicle can limit the maximum speed of the car according to that allowed on the road, keeping it constant, contributing to the driver not receiving a fine for exceeding the maximum speed allowed.

Autopilot helps to reduce the vehicle's fuel consumption. How cruise control works varies according to the model of car and is generally based on the vehicle in front of you, using sensors and cameras to identify its speed and distance. The system is programmed so that the car travels at a speed proportional to the vehicle in front of it or at a speed pre-set by the driver. An ultrasonic sensor is used to measure distance, and its operation is based on an oscillator that emits ultrasonic waves, thus allowing it to detect objects in front of it (THOMAZINI; ALBUQUERQUE. 2005, p.18).

The cruise control is activated by means of buttons located on the steering wheel. There are two ways to set the speed, the first is to accelerate the car to the desired speed and press the cruise control button, and the second is to press the cruise control button after the cruise control mode has been activated to increase or decrease the speed. The system can be disabled in two ways: by pressing the cruise control button or by pressing the accelerator pedal until the pre-set speed is exceeded. In some vehicles, the system is automatically deactivated when the driver firmly steps on the accelerator, brake or clutch pedal (if the car is not automatic) (ROCHA, 2011).

Autopilot doesn't make the car completely autonomous, because as was said at the beginning of this section, the cruise control has the function of assisting the driver, but not replacing him, because it can't identify all dangerous situations or brake in an emergency. With the development of new technologies, systems have been developed so that vehicles are able to read the lanes and stay on course even when cornering. However, this does not remove the obligation to have a driver in the car, so that in the event of an electrical failure, he can take control of the car again (ROCHA, 2011), (DIAS; PEREIRA; PALHARE, 2013).

2.6   ELECTRICAL CIRCUIT COMPONENTS

2.6.1 Transistor

For many years, the valve was used as an amplification device, but in 1947, on December 23rd, scientists William Shockley, Walter H. Brattain and John Bardeen presented the first transistor with an amplification function at Bell Telephone Laboratories. The benefits over the valve were clear: it was smaller and lighter, eliminated the need for heating and avoided the joule effect, had a more robust structure, absorbed less power, was more efficient and worked with lower operating voltages (BOYLESTAD, NASHELSKY, 2014), (ARAÙJO, 2015).

A transistor is a semiconductor component made up of two *n-type* layers and one *p-type layer* or two *p-type layers* and one *n-type layer*. The former is called *npn* and the latter is called *pnp*. The symbols for both are shown in figure 1. The transistor has three terminals: base, collector and emitter. Usually, the abbreviation TBJ, for bipolar junction *transistor,* is applied to this device (BOYLESTAD, NASHELSKY, 2014) (ARAÙJO, 2015).

Figure 1 - Bipolar Junction Transistor (TBJ).

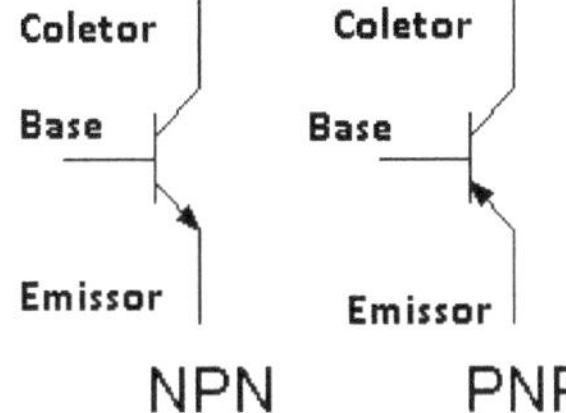

Source: Author.

The operation of a transistor is based on the amplification of currents. When a current is released at the base terminal, a current proportional to the base current is allowed to pass between the emitter and collector terminals. The transistor can work as an electronic switch, but it has to be polarized appropriately: cut-off and saturation. In saturation, the transistor operates as a short (closed switch) between the collector and emitter so that $v_{CE} \cong 0$ V and when it is in cut-off, the transistor works as an open circuit (closed switch) between the collector and emitter, so that $V_{CE} \cong V_{CC}$. At the saturation point (closed switch) the base current is high and at the cut-off point (open switch) the base current is zero (BOYLESTAD, NASHELSKY, 2014).

The use of transistors in electrical circuits can produce a problem when they are switched off, because when the current flow stops, the inductive characteristics of the motor continue to generate current flow (reverse current), which can damage the transistor. To solve this problem, a diode is used in parallel with the motor (ARAÙJO, 2015).

Boylestad and Nashelsky (2013) state that by applying Kirchhoff's current law to the *transistorpnp in* figure 1, considering that it is a single node, we obtain:

$$I_E = I_C + I_B \tag{6}$$

Thus, the emitter current is the sum of the collector and base currents. The cut-off region is defined as that in which the collector current is 0A. In the cut-off region, both junctions of a transistor, base-emitter and base-collector, are reversely polarized. In the saturation region, the base-emitter and base-collector junctions are directly polarized.

For this work, it will be used in all DC analysis of transistor circuits, i.e. if the transistor is in the "on" state, the base-emitter voltage will be considered as follows:

$$V_{BE} = 0,7V \tag{7}$$

Several different circuits are studied and there is a similarity between the analysis of each configuration due to the use of the current of the following important basic relationships of a transistor.

$$I_E = (\beta + 1)I_B = I_C \tag{8}$$

In the analysis of c_I and $_{IB}$ values, they are related by a parameter called beta and expressed by equation 9:

$$\beta = \frac{I_C}{I_B} \tag{9}$$

To make the direct polarization of the base-emitter junction, the base-emitter loop of the *npn* transistor shown in figure 2 is analyzed, applying *Kirchhoff*'s voltage law in the clockwise direction of the loop, equation 10 is obtained.

Figure 2 - Electronic diagram of the power circuit.

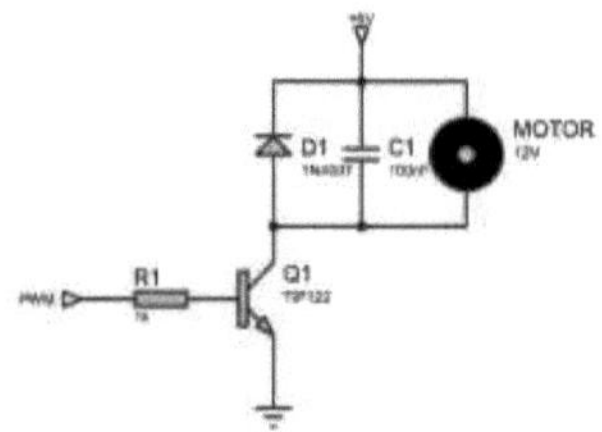

Source: Author.

$$+V_{CC} - I_B R_B - V_{BE} = 0 \tag{10}$$

Isolating I_B in equation 10 gives us

$$I_B = \frac{V_{CC} - V_{BE}}{R_B} \tag{11}$$

For the analysis of the polarization of the collector-emitter section of the transistor, the value of the current is directly related to $_{IB}$ through:

$$I_C = \beta I_B \tag{12}$$

Applying *Kirchhoff*'s law clockwise along the loop of the *npn* transistor gives us

$$V_{CE} + I_C R_C - V_{CC} = 0 \tag{13}$$

Isolating V_{CE} in equation 13 gives us

$$V_{CE} = V_{CC} - I_C R_C \qquad\qquad (14)$$

2.7 DC MOTOR PARAMETERS

2.7.1 Armor Resistance Calculation Method [Ra]

According to Silva (2009), because the armature resistance of a DC motor is very low, it is impossible to measure it accurately using an Ohmmeter. To solve this problem, a direct current is applied to the motor terminals when the motor is blocked. When this happens, the voltage drop as a function of the inductance in the armature winding is zero, as is the induced counter-electromotive force, E, due to the rotor's angular velocity being zero at that moment.

As soon as there is electrical contact between the armature windings and the motor terminals by means of brushes, the resistance seen by these terminals varies according to the angular position of the rotor, when there is a smaller or larger number of turns coming into contact with the brushes. To calculate R_a , a linear adjustment is made to the set of V_a and I_a values measured for different positions (SILVA, 2009).

Silva (2009) states that the linear coefficient of the line resulting from linear regression can be interpreted as the regression error ξ, or as the result of measurement errors. This error can be minimized by applying the Least Squares Method.

$$V_a = I_a R_a \qquad\qquad (15)$$
$$I_a^T V_a = I_a^T I_a R_a \qquad\qquad (16)$$
$$[I_a^T I_a]^{-1} V_a = [I_a^T I_a]^{-1} [I_a^T I_a] R_a = R_a \qquad\qquad (17)$$

2.7.2 *Pulse Width Modulation* (PWM)

The speed of a DC motor can be controlled by PWM (*Pulse Width Modulation*), in which the width of the pulses controls the current supplied to the motor and thus the speed of rotation. As the pulses contain a constant amplitude, they can even start the motor at low speeds or under low power conditions (ARAÙJO, 2015), (GUIMARÂES, 2007).

The PWM principle is based on switching the motor on and off at a fixed frequency, mediated by a switch, usually some form of transistor, causing the motor to rotate at a speed appropriate to the correlation between the related time and the period (T). The correlation between the time a wave is at a high level (vcc) and the complete time is called *Duty Cucle and*, if multiplied by the peak voltage (voltage supplying the motor), an average voltage is obtained which is equivalent to the voltage vdc that would have to be applied to make the motor turn at the same speed. *Duty Cucle* is usually expressed as a percentage (ARAÙJO, 2015).

In this principle, the width of the high pulse is varied while the complete switching time is constant. Figure 2 shows two PWM waveforms, each with a different pulse width (ARAÙJO, 2015).

Figure 3 - Representation of Two PWM Waveforms.

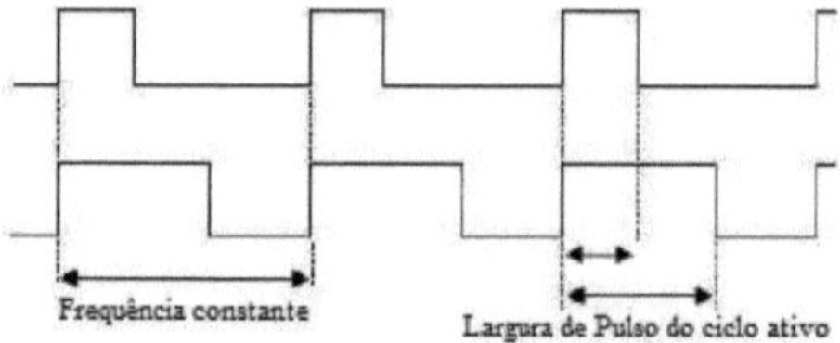

Source: VIVALDINI, 2009.

The average voltage changes depending on how long the signal is at high level (vcc) and how long it is at low level (0V), so you can monitor the intensity of a continuous signal without losing its characteristics. This correlation is called the active cycle, i.e. an active cycle of 33% means that the signal was at high level (vcc) for one third of the time and at low level (0 V) for two thirds of the time. Looking at the theory, the average voltage of a waveform (vdc) is given by equation 18:

$$V_{dc} = \frac{1}{T}\int_0^T V(t)dt \tag{18}$$

Where T is the period of the waveform and V(t) is the time dependence of the voltage. For PWM we have equation 19:

$$V(t) = \begin{cases} V_{pulso}, & 0 \leq t \leq V_p \\ 0, & t_p < t \leq T \end{cases} \tag{19}$$

Where t_p is the pulse duration at logic level and V_{piaso} is the pulse voltage of the PWM signal.

$$V_{dc} = \frac{1}{T}\left(\int_0^{t_p} V_{pulso}\, dt + \int_{t_p}^T 0\, dt \right) \tag{20}$$

$$V_{dc} = \frac{t_p}{T} V_{pulso} \tag{21}$$

The ratio between the pulse width and the wave period is called the *Duty Cycle*. The PWM pulse has a fixed voltage, but its average value varies according to this cycle. The average voltage (*vdc*) is directly proportional to the duty cycle and as this varies between 0 and 1, the average voltage of the wave can vary between 0 and V pulse. It will therefore vary from *vss* to *vdd*, for a voltage variation of 0 to 5 V.

With this procedure, you can vary the average intensity of the current in the motor by feeding it with pulses and controlling their duration, so that the voltage of each pulse remains similar to the maximum voltage of the source, but the average value given to the motor will only be half the input value. To change its speed, simply change the pulses applied (ARAÙJO, 2015), (BRAGA, 2005).

3 DESCRIPTION OF THE PROJECT AND MATERIALS USED

This section describes the prototype electronic accelerator, its materials and the control flowchart.

3.1 GENERAL PROJECT DIAGRAM

This topic will present the block and electrical diagram of the prototype electronic throttle.

3.1.1 Block Diagram

The block diagram of the electronic accelerator prototype is shown in figure 3. It consists of an Arduino® board, sensors, power circuits and actuators.

Figure 4 - Block diagram of the Autopilot prototype.

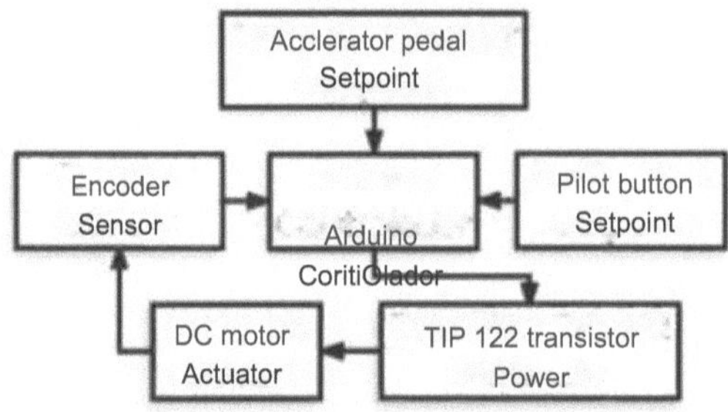

Source: Author.

The design of the autopilot prototype has two *setpoints* (commands), the command to accelerate the engine and the autopilot. The pedal has a spring and gear system which, at the end of its stroke, turns a potentiometer which increases the tension if the accelerator pedal is pressed. Finally, the autopilot button has the function of activating the prototype's automatic mode or disabling it.

Each component in this project is connected to the Arduino® UNO, where it interprets the status of each command (*setpoint)* and activates the power circuit and motor. Thus, by means of the sensor attached to the motor shaft (*Optical Switch/Encoder*), the rotation per minute (RPM) reading is obtained.

3.1.2 Electrical   diagram

Figure 4 shows the electronic diagram of the proposed project's power circuit. This diagram consists of a 1KΩ resistor; a TIP122C transistor whose function is to control the electric current supplied to the motor; a 1N4007 diode whose function is to protect the Arduino® board against reverse voltage generated by the motor's inductance; a 100 nF capacitor whose function is to filter out the noise generated by the motor; and a 12 V DC motor.

Figure 5 - Electronic diagram of the power circuit.

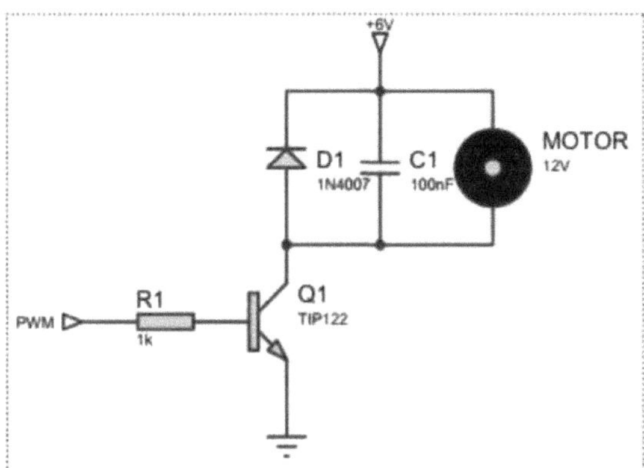

Source: Author.

Figure 5 shows a diagram, which is made up of an Arduino® UNO board, where it reads the *setpoints* and generates the PWM signals to drive the motor. Analog pin A0 connects to the accelerator pedal, digital pin 2 connects to the *optical switch/encoder*, digital pin 3 connects to the pilot button, digital pin 4 connects to the pilot button (+), digital pin 5 connects to the pilot button (-), digital pin 6 is connected to the 12 V motor and digital pins 8 to 13 are connected to the *display*.

Figure 6 - Control diagram.

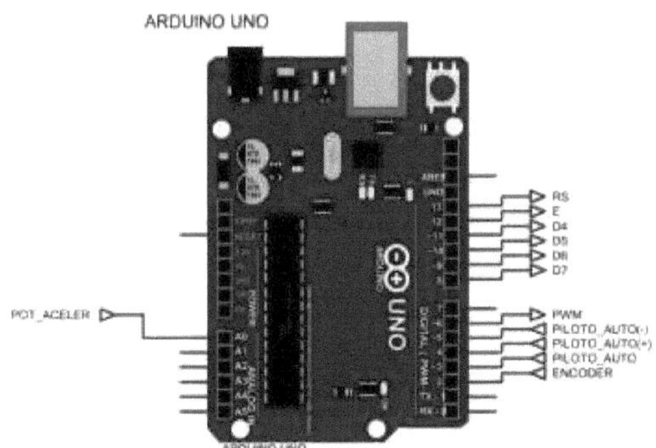

Source: Author.

Figure 6 shows the control diagram, which is made up of 10KΩ potentiometers whose function is to change the voltage supplied to the system, a 1KΩ resistor and the system's activation buttons.

Figure 7 - Command diagram.

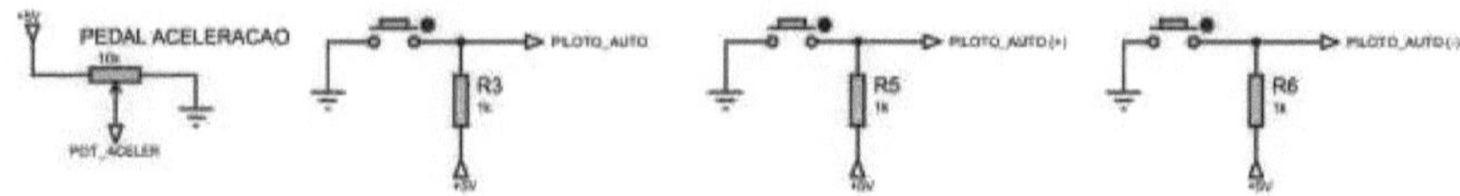

Source: Author.

Figure 7 shows the diagram of the optical key sensor that reads the *encoder* disk, since it has an infrared emitter and receiver at its ends.

Figure 8 - Sensor diagram.

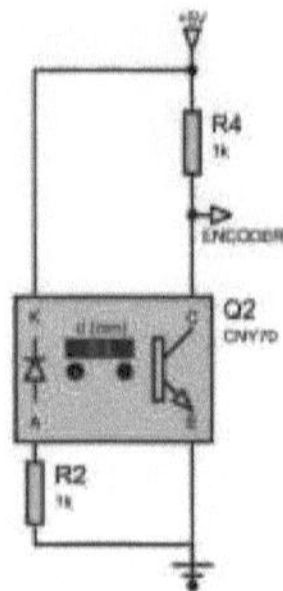

Source: Author.

Figure 8 shows the *display* diagram, which consists of a 16x2 LCD *display,* whose function is to display the motor's rotation speed reading.

Figura 9   - Display diagram.

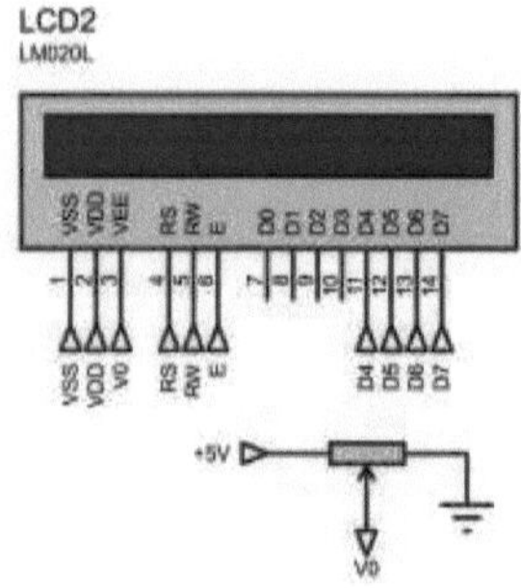

Source: Author.

Figure 9 shows the power supply circuit diagram, consisting of four 1.5 V AA batteries, connected in series, providing a total of 6 V for the entire system.

Figura 10   - Power diagram.

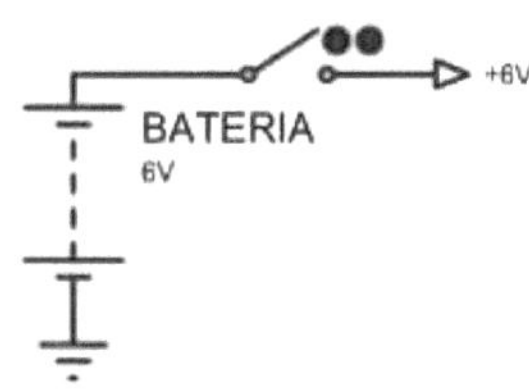

Source: Author.

3.2 PROTOTYPE COMPONENTS

Below are more details of the prototype, the main components and the main reasons for choosing them.

3.2.1 Arduino® UNO

Table 1 shows the characteristics of the Arduino® UNO board, describing the number of analog and digital pins the board has, the microcontroller used, the voltages that can be supplied to the Arduino®, the *clock* speed, *Flash* memory, SRAM, EEPROM and the DC currents of the 3.3 V and I/O pins.

Table 1 - Arduino® UNO features

Microcontroller	ATmega328P
Minimum supply voltage	5V.
Operating voltage (logic level)	5V.
Input voltage (recommended)	7-12V.
Input voltage (limits)	6-20V.
Digital I/O pins	14
Analog input pins	6
DC current per I/O pin	20 mA
DC current to 3.3V pin	50 mA
Flash memory	32 KB (ATmega328P)
SRAM	2 KB (ATmega328P)
EEPROM	1 KB (ATmega328P)
Clock speed	16 MHz

Source: Arduino, 2015.

The Arduino® Uno can be powered via the *Jack* connector with the positive in the center, where the voltage value of the external supply varies from 6-20V, but if powered with a voltage below 7V, the board's operating voltage can fluctuate, and when supplied with a voltage above 12V, the board's voltage regulator can overheat and damage the board. This happens because the voltage regulator dissipates energy in order to input a value and output +5V and +3.3V. For this reason, it is recommended to use external supply voltages ranging from 7-12V.

3.2.2 Prototype cost

Table 2 describes the components used to manufacture the autopilot prototype and the average cost of each material on the virtual market.

Table 2 - Prototype components - Costs.

Description	Reference	QTY	Cost (R$)
Arduino UNO	-	1	35,00
DC motor (12V)	-	1	11,00
Capacitor (Ceramic)	100nF	1	5,00
Diode	1N4007	1	0,10
Protoboard	HK-P50	1	15,00
Resistor	1kΩ	2	0,20
Power meter	10kΩ	3	3,60
LCD *display*	16x1	1	24,40
Sensor *Encoder*	-	1	6,50
Button	-	1	0,30
Battery holder	AA battery	1	5,00
Transistor	TIP122C	1	1,50
Jumper wire	Male	1	5,00
14mm wood	-	1	6,00
Accelerator pedal	Steel	1	30,00
Total			148,60

Source: Author.

Table 2 has been drawn up to give the reader an idea of the cost of manufacturing a prototype electronic accelerator, but some of the materials used in this work can be found in electronic scrapyards.

3.2.3 *Optical key/encoder* sensor

The sensor used in this work, shown in figure 10, is the *optical switch/encoder*, which is of great importance in industrial automation because it provides information on the speed or position of parts that rotate or move in a straight line. The *encoder* consists of a disk assembly with alternating lines and a photoelectric sensor. The photoelectric sensor consists of a diode (LED) and a photodiode or phototransistor. When the disk rotates, electrical pulses are generated by the *encoder*. The position or speed is identified by counting the number of pulses generated and obtaining a frequency. The electrical pulses are sent by the photoelectric sensor to a control circuit as soon as the disk begins to rotate.

Figure 10 - Optical Switch/Encoder Sensor.

Source: Author.

3.2.4 Power Circuits

The DC motors shown in figure 12 run on direct current electricity and are used in a wide variety of areas, such as small robotic projects. To control a motor of this type, you need a higher current than the Arduino® can supply, so in order to amplify this current, you need to set up a circuit with transistors.

In this project, a circuit was used in the common emitter configuration, with the transistor operating as a switch (in saturation and cut-off) for the DC motor, and a diode to protect the Arduino® board, in order to avoid damage caused by reverse currents generated by the motor.

3.2.4.1 TIP121C transistor

The TIP122C transistor, shown in figure 11, is of the *npn* type for medium-power linear switching applications. Some of its characteristics are: I_C = 5 A; V_{CEO} = 100 V; V_{CBO} = 100 V, V_{EBO} = 5 V; hfe = 20 (FAIRCHILD, 2015).

Figure 11 - TIP122C transistor

Source: Author.

3.2.5 Engine

3.2.5.1 DC motor

The motor chosen was a 12 V DC motor with an optical switch and *encoder* disk attached, purchased from a printer.

Figure 12 - DC motor

Source: Author.

3.2.5.2 Determination of reinforcement resistance [Ra]

To determine the armature resistance [R_a of the motor, a bench test was carried out (motor blocking). Table 3 shows the voltage and current values obtained from certain angles using a variable voltage source.

Table 3 - Measurements for determining Ra.

Angle	Va [V]	Ia [A]
0°	1,0	0,025
	1,5	0,038
120°	2,0	0,044
	2,5	0,065
240°	3,0	0,084
	4,0	0,117
	5,0	0,148

Source: Author.

Using the MATLAB *software*, and applying the least squares method, the value of R_a was found to be, $R_a = 35.0433\ \Omega$.

3.2.6 LCD *display*

LCD displays don't have features such as automatic pixel control, they aren't colorful, they don't have active lighting, among other limitations, but they are still widely used in industry. It's enough to see that many cash registers, portable equipment, computer server information systems and others still use this device extensively. Figure 13 shows the *display* used in this work.

Figure 13 - 16x2 LCD *display*

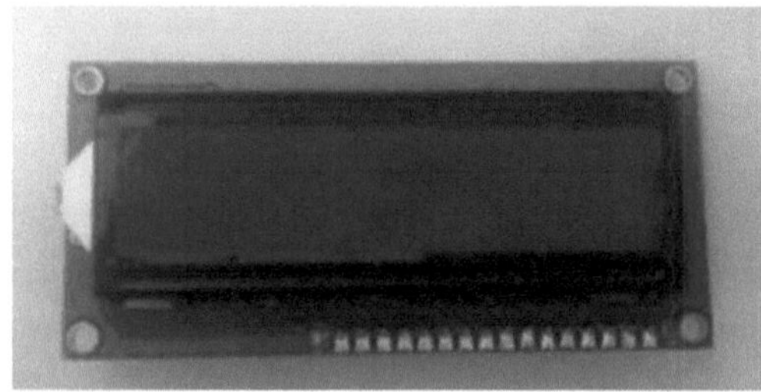

Source: Author.

Table 4 shows the function of each of the 16 pins of the LCD *display* used in the prototype.

Table 4 - Description of connection interface pin.

Pin	Function	Description
1	Food	Ground or GND
2	Food	VCC or +5 V
3	V0	Contrast adjustment voltage
4	RS Selection:	1 - Data, 0 - Instruction
5	R/W selection:	1 - Reading, 0 - Writing
6	And Chip select	1 or (1→ 0) - Enable, 0 - Disable
7	B0 LSB	
8	B1	
9	B2	
10	B3	Data bus
11	B4	
12	B5	
13	B6	

14	B7 MSB	
15	A (if any)	LED backlight anode
16	K (if any)	Cathode for LED backlight

Source: Author.

3.2.7 Physical Structure

This section describes the mechanics of the proposed prototype.

3.2.7.1 Acceleration pedal

The tools and machines used to manufacture the pedal were: a grinder to cut the round bars, plates and provide a final finish, a lathe to manufacture the axles, slide bearings and springs, a welding machine to connect the components such as the bearings, base and pedal plate, a drill to make some holes in the potentiometer base and pedal fixing base; a mechanical milling machine and milling module to manufacture the gears.

The materials used were a SAE 1020 carbon steel sheet with a smooth finish and trimmed edges, with a thickness of 3.17 mm and dimensions of 1000x2000 mm; a galvanized sheet of SAE 1020 carbon steel, with a smooth finish and untrimmed edges, thickness 0.95 mm, dimensions 1000x3000 mm, density 8.07 kg/m^2 , and a cylindrical bar of drawn AISI 304 steel, raw, with a section dimension of 12.70 mm, nominal weight 0.99 kg/m, a cylindrical bar, of SAE 1020 alloy steel, rolled, raw, with a section dimension of 7.94 mm and nominal weight 0.39 kg/m.

Also used was a cylindrical bar of AISI 304 steel, rolled, blanks, with a section dimension of 15.88 mm and a nominal weight of 1.55 kg/m. Coated carbon steel electrode, classification E6013, rutile coating, alternating/continuous current, positive polarity, core diameter 2.50 mm, length 350 mm, ESAB OK 46.00. Two cylindrical helical springs, traction loading type, made of spring steel, uncoated, right-hand torsion type, flat end, wire diameter 1.00 mm, with internal diameter 8.00 mm and free length 27.00 mm.

Figure 14 shows an image of the accelerator pedal.

Figure 14 - Acceleration pedal

3.3 AUTOPILOT OPERATION

To deactivate the cruise control, the driver simply presses the pilot button and the system is quickly switched off, giving full control of the car back to the driver. When the system is activated, it can increase or decrease the pre-set speed, making it more comfortable for drivers. To increase the speed, the driver presses the button in the direction of the "+" symbol and to decrease the speed, the driver presses the button in the direction of the "-" symbol.

3.4 system control flowchart

Figure 15 shows a flowchart to help visualize and understand how the electronic throttle control works.

The program starts by declaring the variables and configuring the I/O pins: defining the acceleration pedal as the analog input, the autopilot as the digital input, and the PWM signal on the DC motor as the output. The program then enters an "infinite loop" where it reads the pedal to enable the PWM to drive the motor speed as the pedal angle varies. If the autopilot is enabled, it will enter the automatic routine, which will allow the user to increase or decrease the motor speed.

Figure 15 - Flowchart of the Proposed System

Início
Declaração de Variáveis
Configuração das portas I/O
Pedal e Piloto como entrada analógica
Habilitar PWM
Leitura do Pedal
Pedal Aceleração?
Não
Sim
Acelerar motor
Velocidade momentânea
Botão Piloto?
Não
sim
Habilita automático
Rotina automático atuador?
Não
Sim
Botão Piloto?
Não
Pedal Acelerador?
Não
Piloto -?
Não
Piloto +?
Não
Sim
Sim
Sim
Sim
Desativar automático
Acelerar motor
Diminuir velocidade
Aumentar velocidade
Desacelera motor
Velocidade momentânea

4 RESULTS

To obtain the results of this work, experiments were carried out to generate graphs showing the performance of the prototype. To measure this experiment, a Minipa model MO-2100D digital oscilloscope was used to collect the motor's frequency, f, and period, T, data.

4.1 WAVE SHAPE

Figure 16 shows the waveform of the motor's PWM signal, showing a DC motor duty cycle of 50% of the supply voltage.

Figure 16 - Waveform of the PWM Signal from the DC Motor Signal.

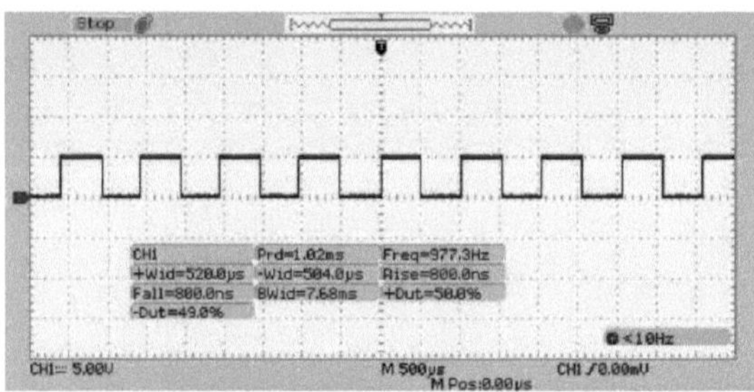

Source: Author.

4.2 SENSOR TEST

The speed equation in RPM is:

$$RPM = f \frac{60}{32} \tag{20}$$

Where RPM is the rotation per minute, f *is* the frequency, 60 is the number of seconds in a minute, and 32 is the number of blades on the *encoder* disk.

Experiments were carried out to measure the voltage variation when the accelerator pedal was pressed. At this stage, it was observed that the minimum voltage for the DC motor to work is 1.5 V. This is due to the characteristics of DC motors, which require a significant voltage to operate. Thus, a table was drawn up with the performance data of the motor used in the prototype. In this table, we looked at the variation in period and frequency when the voltage is varied, thus obtaining the RPM values using equation 20.

Table 5 - Measurements to determine motor period and frequency

Voltage (V)	Period (ms)	Frequency (Hz)	Speed (RPM)
1,500	3,570	280,100	662,060

1,700	2,500	399,600	762,190
1,900	2,330	429,900	924,560
2,100	1,920	520,800	1080,000
2,300	1,670	597,300	1186,690
2,500	1,490	671,100	1346,810
2,700	1,350	741,800	1469,440
2,900	1,220	818,300	1621,880
3,100	1,100	907,400	1754,810
3,300	1,000	996,000	1893,750
3,500	0,940	1060,000	2081,250
3,700	0,874	1140.000	2193,750
3,900	0,816	1230,000	2362,500
4,100	0,784	1280,000	2512,500
4,300	0,728	1370,000	2662,500
4,500	0,690	1450,000	2793,750
4,700	0,654	1530,000	2962,500
4,900	0,620	1610,000	3093,750
5,100	0,594	1680,000	3262,500

Source: Author.

4.3 RELATIONSHIP BETWEEN SPEED AND VOLTAGE

Using the data in Table 5, it was possible to find out the speed in RPM and generate a graph of the relationship between speed (RPM) and voltage (V), using MATLAB *software*. Figure 16 shows the graph of the relationship between speed (RPM) and voltage (V).

Figure 17 - Graph of the Ratio between Speed (RPM) Versus Voltage (V)

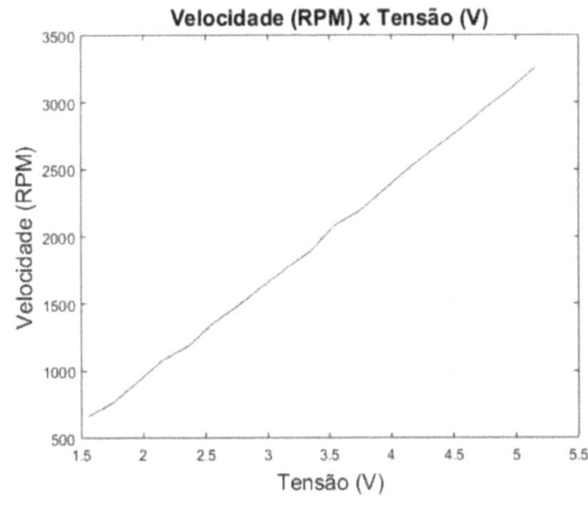

Source: Author.

31

4.4 RELATIONSHIP BETWEEN VOLTAGE AND ANGLE

Using trigonometry concepts, it was possible to obtain the value of each angle when the accelerator pedal is pressed. Using a multimeter, it was possible to obtain the voltage value for each angle variation. A table was drawn up relating the variation in voltage generated to the variation in pedal angle. It was observed that the greater the angle, the greater the voltage generated. Table 6 shows the data collected.

Table 6 - Measurements to determine motor period and frequency

Voltage (V)	Angle (Degrees)
0,00	0.00
0,51	4,60
1,36	9,71
2,21	15,30
3,39	21,34
4,30	27,74

Source: Author.

With the data from table 6 *already* collected, MATLAB *software* was used to generate the curve representing the relationship between voltage and angle. Figure 17 shows the curve representing the relationship between voltage and angle.

Figure 18 - Graph of the Relationship Between Voltage (V) Versus Angle (°)

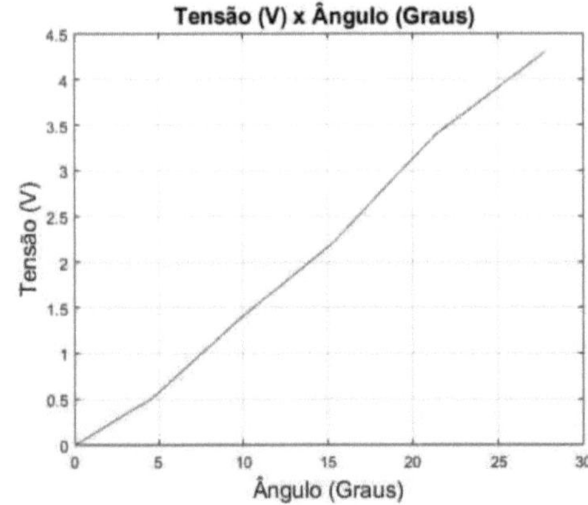

Source: Author.

4.5 COST ANALYSIS AND FEASIBILITY OF THE PROTOTYPE

Table 2 of this work shows that the total cost of the prototype was 148.60 reais. A market survey was carried out on the prices of autopilots for cars and it was found that the average price of an autopilot is 1,800.00 reais. By using low-cost materials, the prototype had significantly reduced costs compared to a commercially available speed controller model. Figures 19 and 20 show the top and front views, respectively, of the completed prototype.

Figure 19 - Top view of the prototype (°)

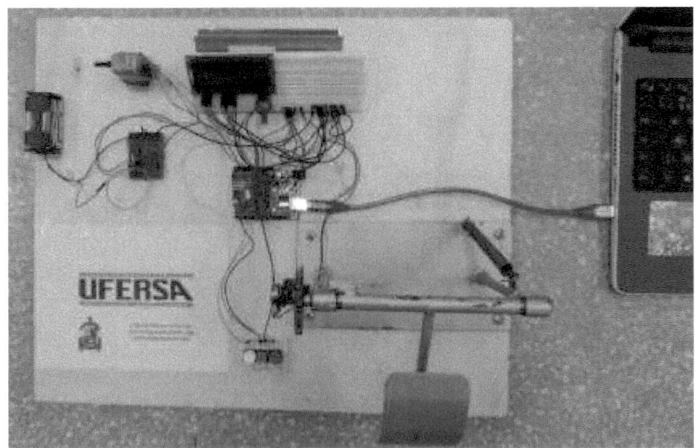

Source: Author.

Figure 20 - Prototype front view (°)

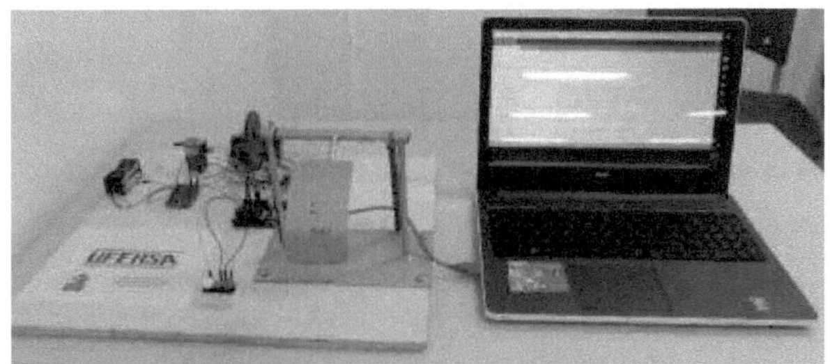

Source: Author.

5   CONCLUSION

The mechanical manufacture of the pedal was satisfactory, as it met the requirement of providing the relationship between the angular displacement and the rotation of the *encoder*. Thus, its design was conceived and elaborated in such a way as to also promote a similarity with the pedals sold by the main car manufacturers.

The three-dimensional modeling *software* Autodesk Inventor was used to design the pedal by making various parts of it, such as the wooden base, the pedal base, the pedal shaft, the potentiometer shaft, the pedal plate and the pedal rod. In addition, the screws, gears and spring were imported from the *software* library. In this way, the assembly of the system served as a visualization for the manufacturing process, thus facilitating its manufacture as shown in the appendix.

The circuit proposed for the work was made with an open loop due to the system not being monitored in real time, so the electrical design was simplified, allowing for less computational cost on the Arduino® *software* platform, while still meeting the proposed objectives.

The prototype was controlled using code implemented in the Arduino® language based on C/C++ on the Arduino® platform, and then inserted into the *hardware*. It was therefore possible to implement routines for the electronic throttle to make it possible to represent the angular position of the throttle with the rotation of the tachometer. In addition, the autopilot was implemented by writing subroutines that were called in the main program in order to activate the speed controller, which, once activated, is responsible for increasing or decreasing the system's speed. While still in operation, the autopilot is deactivated when the speed reported by the mechanical pedal is higher than the speed set when the autopilot is activated.

The graph of the relationship between speed and voltage was approximately linear, as was that of voltage and angle, indicating that it was possible to mathematically model the system using an affine function. However, the tachometer only started moving from an angular position of 11 degrees, indicating that there is a minimum voltage to overcome the motor's inertia.

The cost survey showed the financial viability of the prototype, which is comparatively more viable than commercial models because it uses electronic materials that are cheap on the market. It is therefore possible to use the procedures developed in this work to formulate electronic accelerators that are competitive with existing ones. With this idea in mind, the final cost can be reduced by using cheaper electronic components than those used to make the prototype.

The pedal was protected against corrosion using one of the most widely used procedures in the mechanical industry: painting. Therefore, the steel components were sanded and then a thin layer of paint was applied using the spray painting procedure. This procedure also gave the system a better visual presentation.

For future work, it is proposed to develop a braking system and implement it in the system already designed. Another proposal would be to design a closed-loop system, which would allow greater control of the system.

6   REFERENCES

AFONSO, Antonio Pereira; FILONI, Enio. Electronics: electrical circuits. **Revista Fundaçâo Padre Anchieta**, v.1, Sao Paulo, 2011.

AGUIA, Iana Assunçao de; PASSOS, Elizete. Technology as a path to citizen education. **Cairu Magazine,** 2014.

ALMEIDA, Danilo Alves; LUZ, Larissa Pavarini da. Access control system for condominiums using arduino technology**, Revista E-f@tec**, Garça, v.4, n.1, 2014.

ARAÙJO, Patricia Rodrigues de. **Development of a robot for cleaning tasks.** 2015. TCC (Monograph in Science and Technology) - Universidade Federal Rural do Semi-Arido, Caraùbas, 2015.

BESSON, Thomas et al. Low cost venom extractor based on arduino® board for electrical venom extraction from arthropods and other small animals. **Revista Toxicon**, v.118, p. 156161, August 2016.

BOYLESTAD, R. L; NASHELSKY, L. **Electronic Devices and Circuit Theory**. 8ª Ed. Sao Paulo: Prentice Hall, 2014.

BRAGA, Newton. **Basic Electronics for Mechatronics**. 2005 1ª Ed. Sao Paulo: EditoraSaber, 2005. 160p.

DIAS, Jullierme Emiliano Alves; PEREIRA, Guilherme Augusto Silva; PALHARES, Reinaldo Martinez. **Speed control of an autonomous car.** Article (Postgraduate Program in Electrical Engineering) - Federal University of Minas Gerais, Belo Horizonte, 2013.

EVANS, Martin; NOBLE, Joshua; HOCHENBAUM, Jordan. **Arduino in Action**. 1ª Ed. Sao Paulo: Novatec, 2013. 424 P

FAIRCHILD. *Datasheet* **TIP41A/ TIP41B/ TIP41C NPN Epitaxial Silicon Transistor.**

Handout, November 2014.

FRANCESCONI, Tiago. Theoretical technical drawing. Workbook - **PRODUTRÔNICA**, 2010.

FRAUCHES-SANTOS, C.; ALBUQUERQUE, M. A.; Oliveira, M. C. C.; ECHEVARRIA, A. A corrosão e os agentes anticorrosivos, **Revista virtual de Quimica,** Seropédica, v.6, n.2, p. 293-309, dezembro 2013.

FREITAS, Elias J. de R. et al. **Development of embedded automation for a mobile robot based on a passenger car.** Article (Research Group for the Development of Autonomous Vehicles) - Federal University of Minas Gerais, 2009.

GUIMARAES, Fàbio de Almeida. **Development of a Mobile Robot used to explore hostile environments.** Dissertation (Master's Degree in Chemical and Biochemical Process Engineering) - Maua Engineering School of the Maua Institute of Technology University Center, Sao Caetano do Sul, 2007.

HIBBELER, R. C.. **Dynamics**: mechanics for engineering, 12^a Ed. Sao Paulo: Pearson Prentice Hall, 2011. 591 p.

KELMANN, Gizela; BERNARDO, Wanderley Marques. Is the electroencephalogram with brain mapping superior to the traditional electroencephalogram in the diagnosis of neurological pathologies?, **Revista SciELOBrasil**, v.58. n.1, Janeiro/fev. 2012.

KUPSCH, Daniela C. C.. **Spial:** a learning support tool for software process improvement.223 f. 2012. Thesis (Doctorate in Computer Science) - Federal University of Minas Gerais, Belo Horizonte, 2012.

LASKARA, M. Rahaman et al. Weather forecasting using arduino based cube-sat. **Procedia Computer Science Magazine**, v.89, p. 320-323, 2016.

LUZ, A. P.; RIBEIRO, S.; PANDOLFELLI, V. C.. Using wettability to investigate the corrosion behavior of refractory materials, **Revista ResearchGate,** Sao Carlos, p. 174-183, 2008.

MORAES, Maria Beatriz dos Santos A.; RIBEIRO-TEIXEIRA, Rejane M.. **Electric circuits:** new and old technologies as facilitators of meaningful learning at secondary level. Dissertation (Professional Master's Degree in Physics Teaching) - Institute of Physics - Federal University of Rio Grande do Sul, Porto Alegre, v.17, n.1, 2006.

MOURA, Adolpho Marlon Antoniol de. Laboratory rodent nutrition: paradigms and challenges. **Revista BVS-Vet**, v.2, n.4, 2014.

MUNAWAR, Usman et al. Low cost wireless sensor network based intelligent retina controlled computer. **Procedia Engineering Journal**, 2015. v.107, p. 366-371, 2015.

NGUYEN, Katrina P. et al. Feeding experimentation device (fed): a flexible open-source device for measuring feeding behavior. **PubLMed journal- ncbi.nlm.nih.gov,** 2016.

PANDEY, Anish et al. Mobile robot navigation in unknown static environments using amphis controller. **Perspectives in Science Journal**, v.8, p.421-423, September 2016.

PERTENCE, Antônio E. M.; SANTOS, Daniel M. C.; JARDIM, Helton Vilela. Development of didactic models for teaching mechanical design using the concept of rapid prototyping, Article - **COBENGE,** Porto Alegre, 2001.

PONCE, C.R. et al. Automated chair-training of rhesus macaques. **Journal of Neuroscience**

Methods. v.263, p. 75-80, April 2016.

ROCHA, Tiago Nunes da. **Autopilot for control and navigation maneuvers of the AtlasCar**. Dissertation (Master's Degree in Mechanical Engineering) - University of Aveiro, Aveiro, 2011.

S. Grassini et al. A simple arduino-based eis system for in situ corrosion monitoring of metallic works of art. **Revista Measurement**, Torino, v.89, July 2016.

SAEED, Adil; KHAN, Zulfigar A.; NAZIR, Mian Hammad. Times dependent surface corrosion analysis and modelling of automotive steel under a simplistic model f variations in environmental parameters, **Journal of Materials Chemistry and Physics,** Bournemouth, v.178, p. 65-73, May 2016.

SHAMES, Irving H.. **Dynamics**: mechanics for engineering, 4^a Ed. Sao Paulo: Prentice Hall, 2003. 632p.

SILVA, Tiago Mendonça da. **Automatic Control of the Braking Mechanism of an Unmanned Vehicle.** 2009. TCC (Degree in Control and Automation Engineering) - Federal University of Minas Gerais, Belo Horizonte, 2009.

SOARES, Claudio Cesar Pinto. A historical and scientific approach to graphic representation techniques, **GRAPHICA,** Curitiba, 2007.

THOMAZINI, Daniel.; ALBUQUERQUE, Pedro U. B. **Sensores Industriais - Fundamentos e Aplicações**. 5^a Ed. Sao Paulo: Èrica, 2005. 222p.

YAPICI, Murat; KOLDEMIRB, Birsen. Developing Innovative Applications of Technical Drawing Couser at the Maritime Education, **Procedia Journal - Social and Behaviorial Sciences**, Istanbul, v.195, p. 2813-2821, July 2015.

YU, Je-hun; SIM, Kwee-bo. Classification of color imagination using emotiv epoc and event-related potential in electroencephalogram. **Optik - International Journal for Light and Electron Optics**, v.127, p. 9711-9718, October 2016.

APPENDIX 1 - PROTOTYPE'S GENERAL ELECTRICAL DIAGRAM.

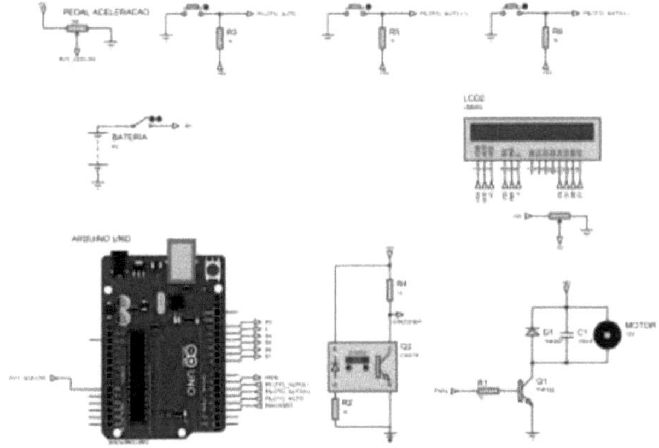

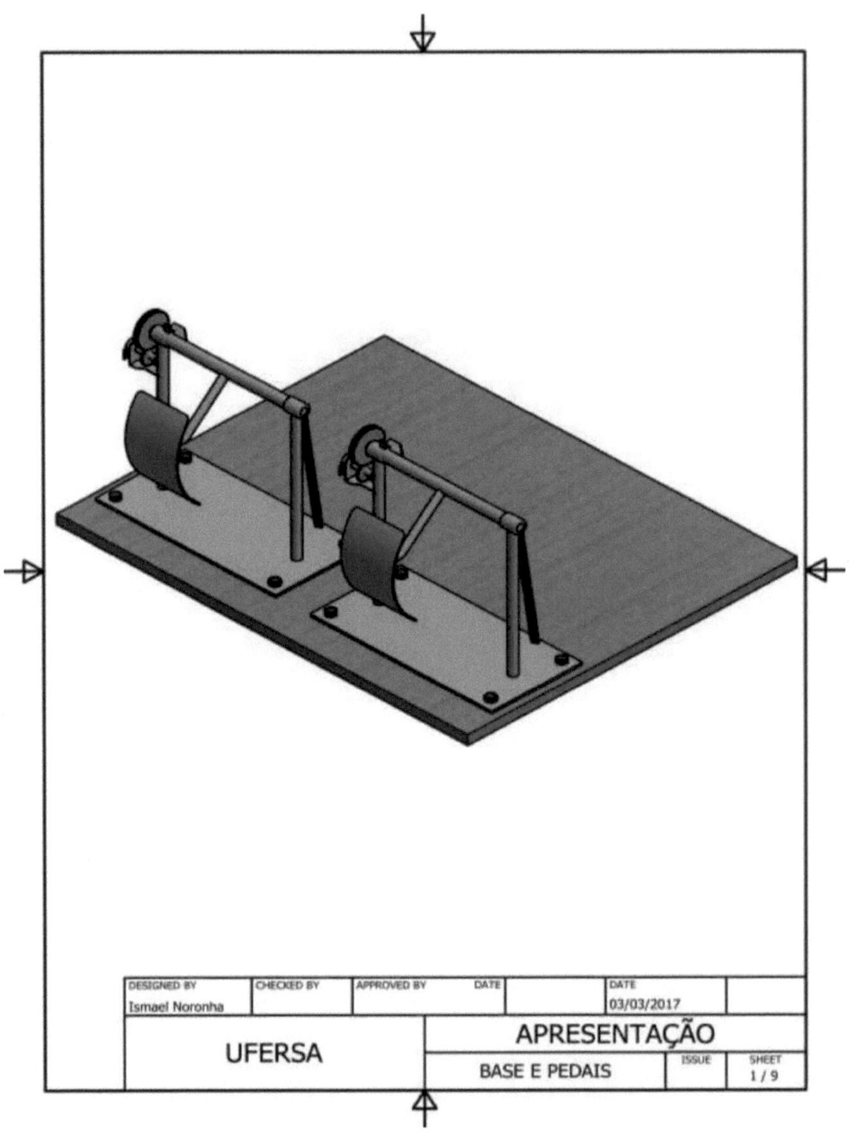

DESIGNED BY
Ismael Noronha
CHECKED BY
APPROVED BY
DATE
DATE
03/03/2017
UFERSA
APRESENTAÇÃO
BASE E PEDAIS
ISSUE
SHEET
1 / 9

APPENDIX 3 - Main views.

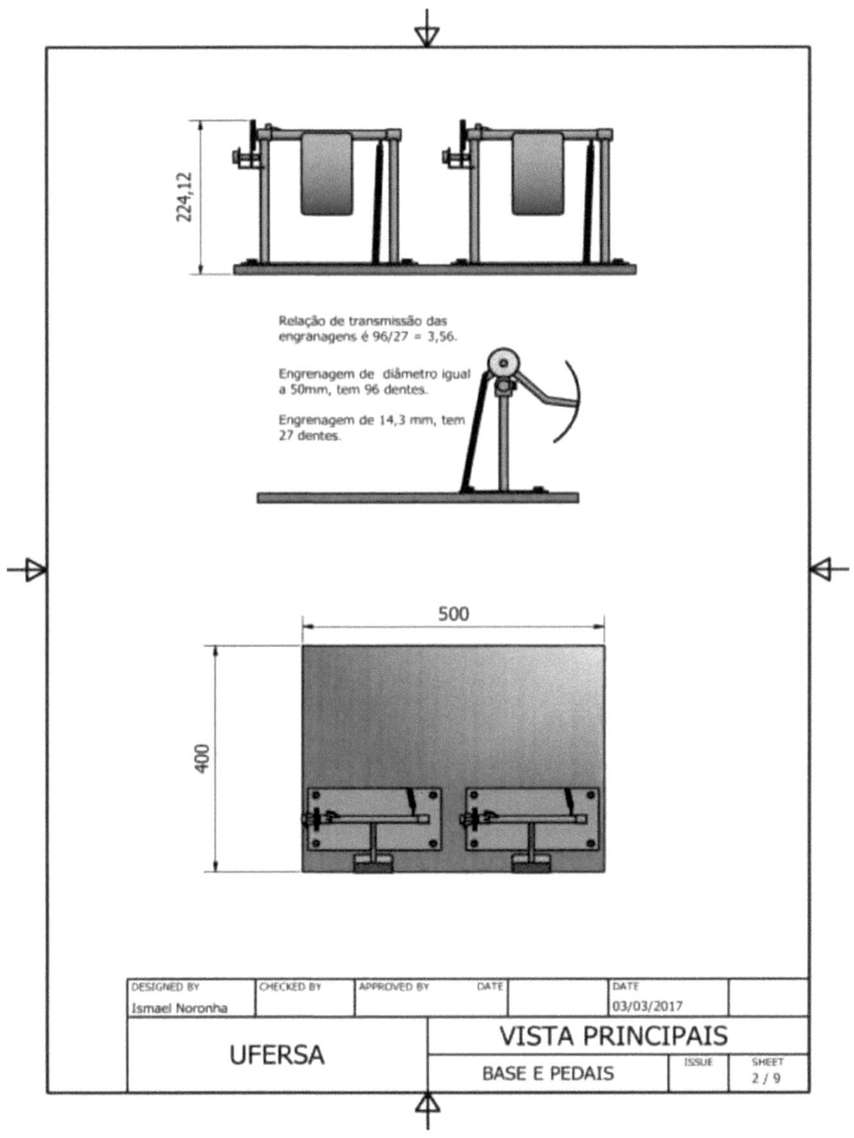

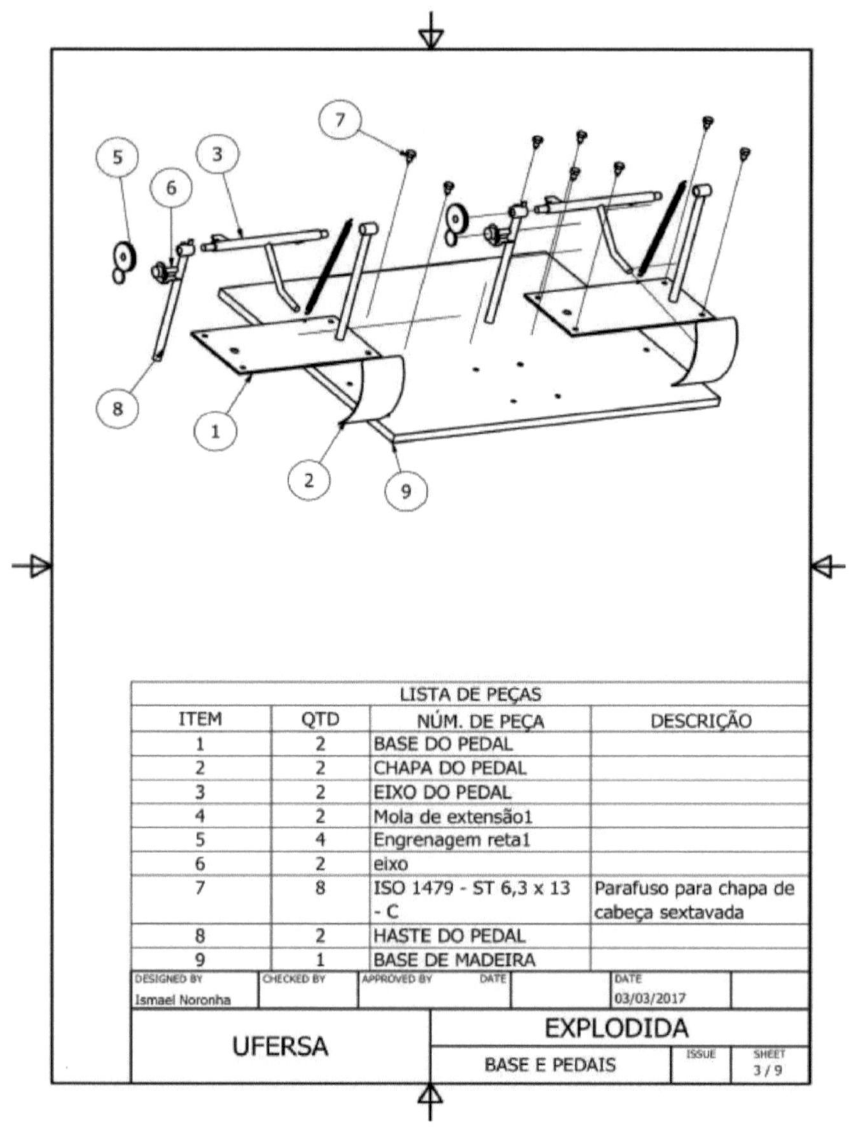

LISTA DE PEÇAS			
ITEM	QTD	NÚM. DE PEÇA	DESCRIÇÃO
1	2	BASE DO PEDAL	
2	2	CHAPA DO PEDAL	
3	2	EIXO DO PEDAL	
4	2	Mola de extensão1	
5	4	Engrenagem reta1	
6	2	eixo	
7	8	ISO 1479 - ST 6,3 x 13 - C	Parafuso para chapa de cabeça sextavada
8	2	HASTE DO PEDAL	
9	1	BASE DE MADEIRA	

DESIGNED BY	CHECKED BY	APPROVED BY	DATE		DATE	
Ismael Noronha					03/03/2017	

UFERSA	EXPLODIDA		
	BASE E PEDAIS	ISSUE	SHEET 3 / 9

APPENDIX 5 - Pedal base.

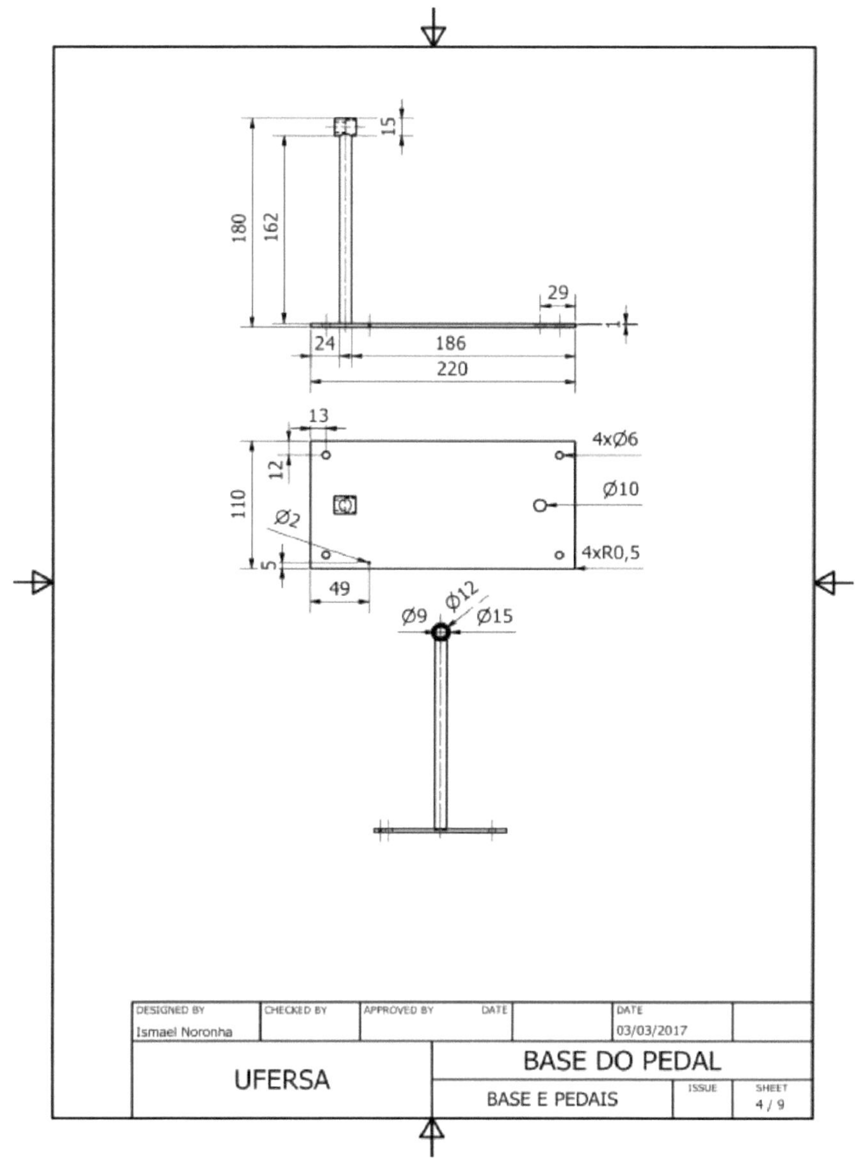

APPENDIX 6 - Pedal rod.

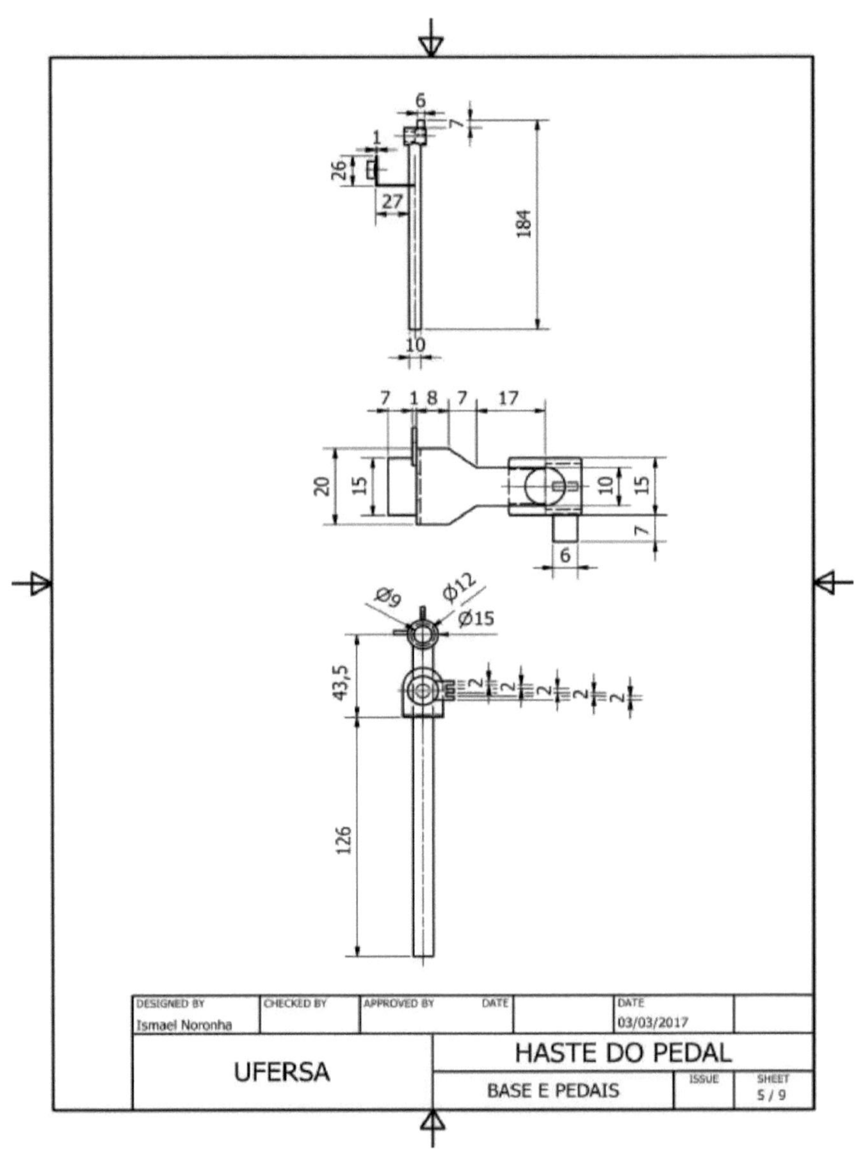

3,18
2xR2
10
R1
4,52
2
65,75
R9,05
72,7°
97,3°
37,87
2,2
R0,95
9,05
Ø8,1
76,95
76,95
2,86 9,31
2,2
0,4
12,2
6,45
Ø12,3 Ø9,25
Ø8,75
11
162
15
188
DESIGNED BY
Ismael Noronha
CHECKED BY
APPROVED BY
DATE
DATE
03/03/2017
UFERSA
EIXO DO PEDAL
BASE E PEDAIS
ISSUE
SHEET
6 / 9

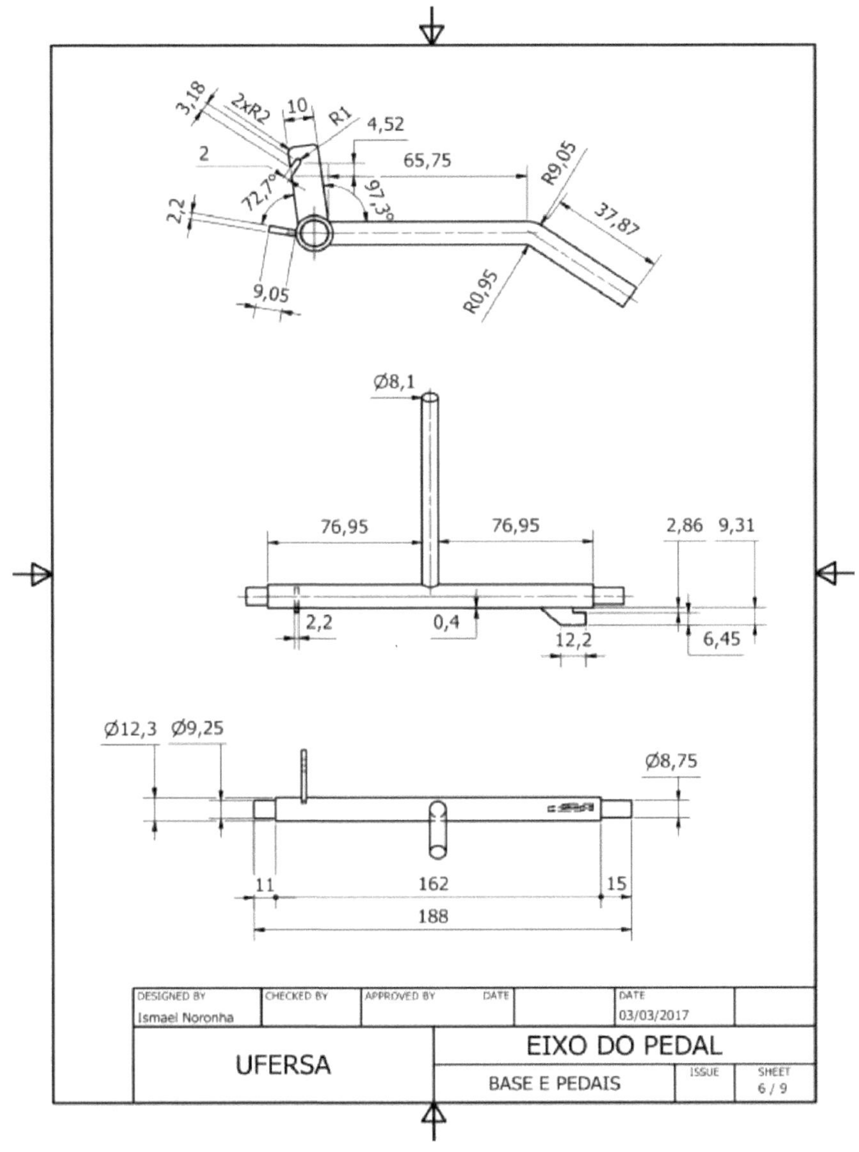

45

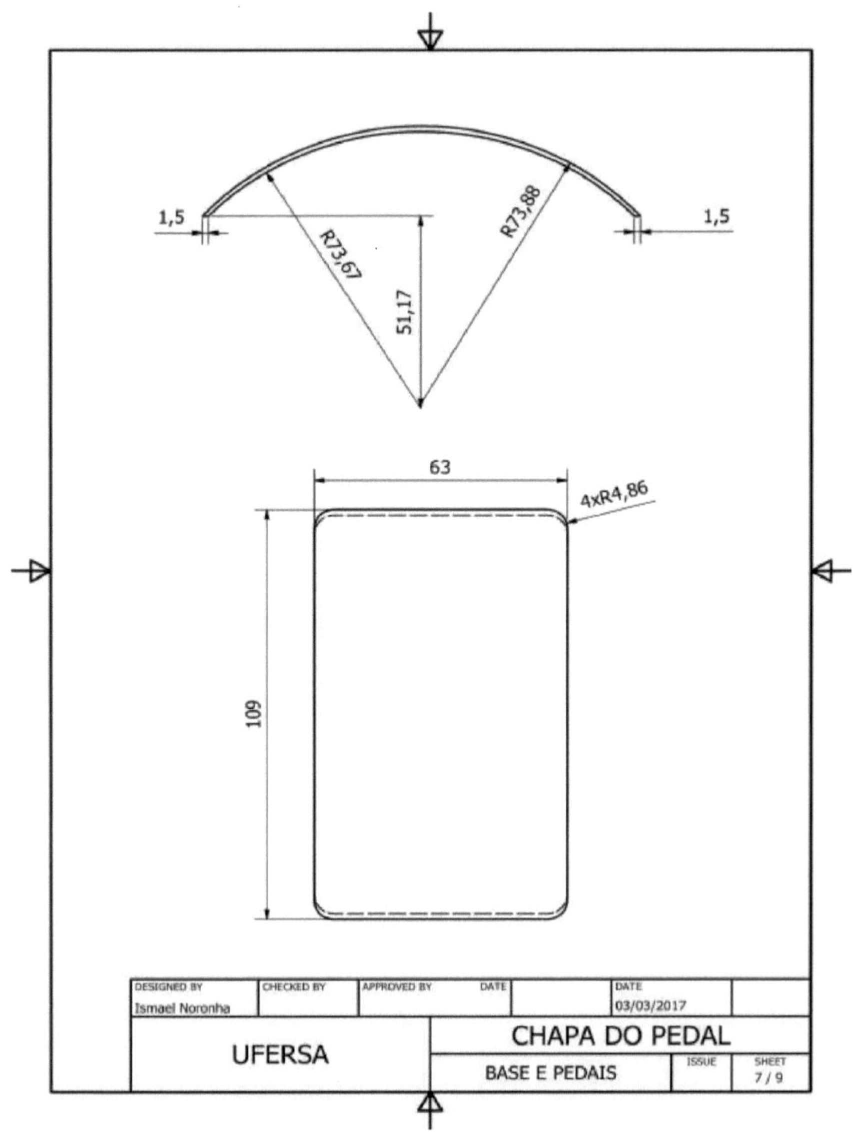

1,5
1,5
R73,67
R73,88
51,17
63
4xR4,86
109
DESIGNED BY
Ismael Noronha
CHECKED BY
APPROVED BY
DATE
DATE
03/03/2017
UFERSA
CHAPA DO PEDAL
BASE E PEDAIS
ISSUE
SHEET
7 / 9

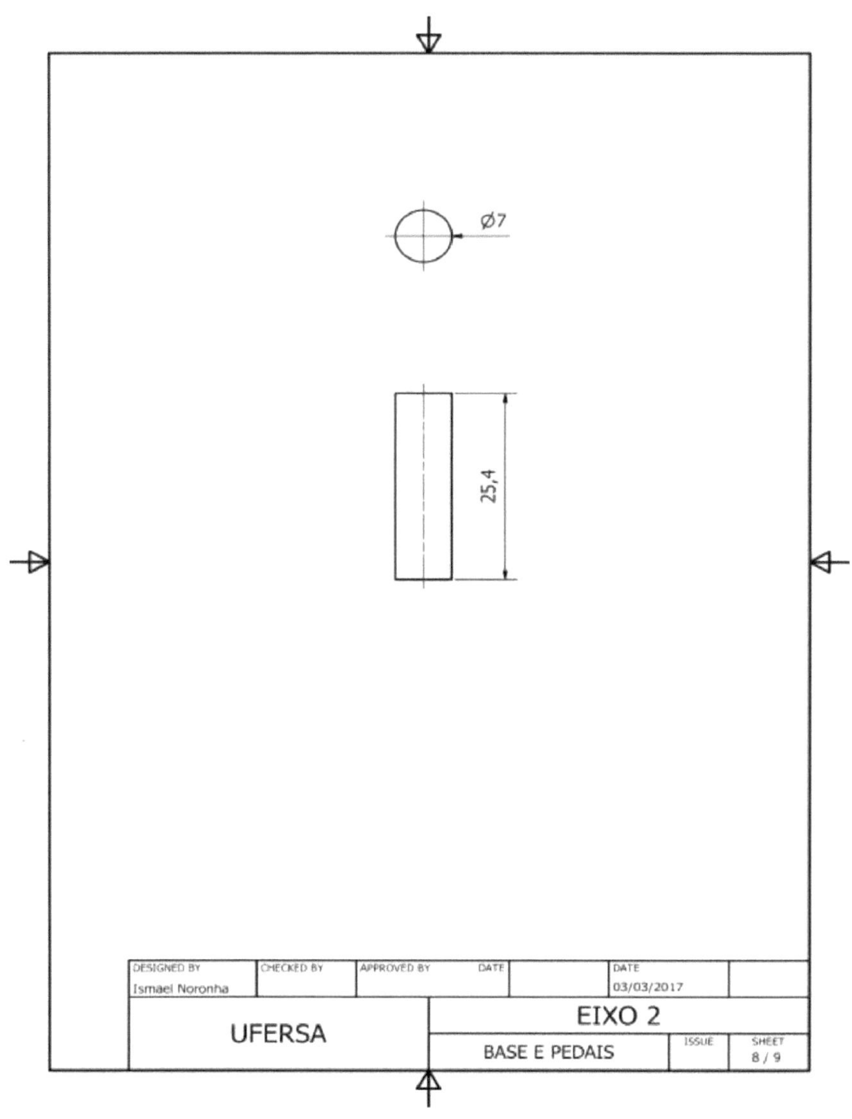
Ø7
25,4
DESIGNED BY
Ismael Noronha
CHECKED BY
APPROVED BY
DATE
DATE
03/03/2017
UFERSA
EIXO 2
BASE E PEDAIS
ISSUE
SHEET
8 / 9

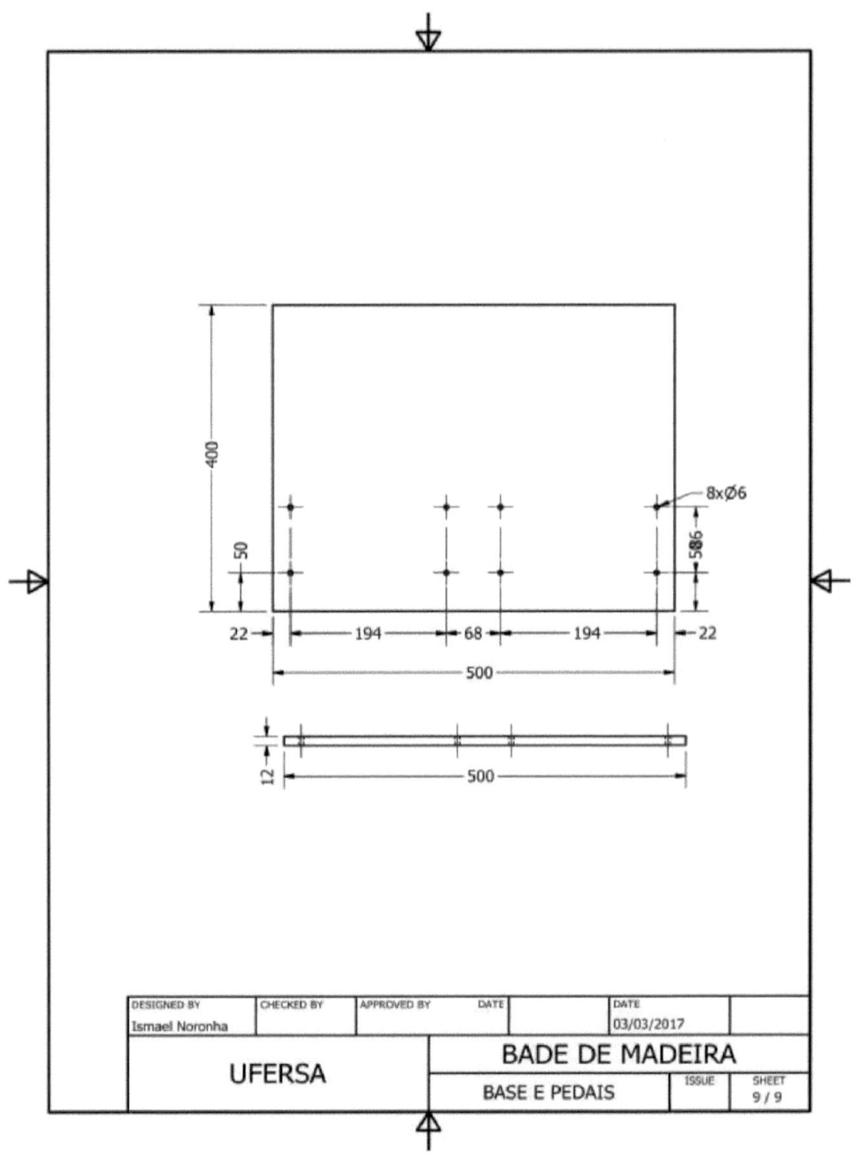
400
50
22
194
68
194
22
500
8xØ6
585
12
500
DESIGNED BY
Ismael Noronha
CHECKED BY
APPROVED BY
DATE
DATE
03/03/2017
UFERSA
BADE DE MADEIRA
BASE E PEDAIS
ISSUE
SHEET
9 / 9

APPENDIX 11 - Prototype program code.

```cpp
//..............................//
//          Bibliotecas        //
//..............................//

#include <LiquidCrystal.h>

//.........................................//
//   Rotina de Declaração das Variaveis    //
//.........................................//

const int pot_aceler = A0;
const int pot_freio = A1;
const int encoder = 2;
const int piloto_auto = 3;
int botao_incrementa = 4;
int botao_decrementa = 5;
const int PWM_motor = 6;
LiquidCrystal lcd(13, 12, 11, 10, 9, 8);

int val_pot_aceler = 0;
int val_pot_freio = 0;
int val_piloto_auto = 0;
int val_aux_piloto =0;

volatile long rpmcont = 0;
long int rpm = 0;
unsigned long tempo = 0;
int porcento = 0;

int piloto = 0;
int val_botao_incrementa = 0;
int val_botao_decrementa = 0;
int aux_piloto = 0;
int aux2_piloto = 0;
int aux3_piloto = 0;
char statu;

//.........................................//
//          Rotina de Configuração         //
//.........................................//

void setup()
 {
 Serial.begin(9600);
 lcd.begin(16, 2);
```

```cpp
attachInterrupt(digitalPinToInterrupt(2), funcao_rpm, FALLING);
pinMode(pot_aceler, INPUT);
pinMode(pot_freio, INPUT);
pinMode(piloto_auto, INPUT);
pinMode(encoder, INPUT);
pinMode(PWM_motor, OUTPUT);
lcd.clear();
lcd.print("   TCC - C&T   ");
lcd.setCursor(0, 1);
lcd.print(" Ismael Noronha ");
delay(1000);
digitalWrite(piloto_auto,HIGH);
digitalWrite(botao_incrementa,HIGH);
digitalWrite(botao_decrementa,HIGH);
}

//.................................................//
//                RPM                    //
//.................................................//

void funcao_rpm()
 {
 rpmcont++;
 }

//.................................................//
//          Display LCD            //
//.................................................//

void display_LCD()
 {
 lcd.clear();
 lcd.setCursor(0,0);
 lcd.print("___TACOMETRO____");
 lcd.setCursor(0,1);
 lcd.print(rpm);
 lcd.print ( "RPM");
 lcd.print ( "  ");
 lcd.print (porcento);
 lcd.print ( "% ");
 lcd.print (statu);
 delay(100);
 }

//.................................................//
//            Leitura             //
//.................................................//

void leitura()
 {
```

```
 val_pot_aceler = analogRead(pot_aceler)/4;
 val_pot_freio = analogRead(pot_freio)/4;
 val_piloto_auto = digitalRead(piloto_auto);
 }


//.....................................................//
//          Rotina Principal           //
//.....................................................//

void loop()
{

if(millis()- tempo >= 1000){
detachInterrupt(digitalPinToInterrupt(2));
rpm = (rpmcont *(60/32));
rpm = rpm*2.1;
rpmcont = 0;
tempo = millis();
display_LCD();
attachInterrupt(0, funcao_rpm, FALLING);
}

leitura();

if((val_pot_aceler>0)&&(aux_piloto==0)){
statu = 'd';
porcento = (val_pot_aceler*100)/255;
analogWrite(PWM_motor, val_pot_aceler);
val_aux_piloto=val_pot_aceler;
display_LCD();
}

leitura();

if(val_pot_freio>50){
porcento = (val_pot_freio*100)/255;
analogWrite(PWM_motor, val_pot_freio);
display_LCD();
}

if((val_piloto_auto == LOW)&&(aux2_piloto==0)){
aux_piloto = 1;
aux2_piloto = 1;
}

if((aux_piloto==1)&&(val_piloto_auto==HIGH)){
statu = 'a';
piloto_automatico();
}
```

```cpp
if((val_piloto_auto == LOW)&&(aux2_piloto==2)){
aux3_piloto=3;
}
if ((val_piloto_auto == HIGH)&&(aux3_piloto==3)){
statu = 'd';
aux_piloto = 0;
aux2_piloto = 0;
aux3_piloto = 0;
leitura();
porcento = (val_pot_aceler*100)/255;
analogWrite(PWM_motor, val_pot_aceler);
val_aux_piloto=val_pot_aceler;
display_LCD();
}
}


//...................................................... ....//
//          Piloto Automatico            //
//......................................................//

void piloto_automatico()
{
aux2_piloto=2;
val_botao_incrementa = digitalRead(botao_incrementa);
val_botao_decrementa = digitalRead(botao_decrementa);

if(val_botao_incrementa == LOW){
val_aux_piloto=val_aux_piloto + 10;
}
if(val_botao_decrementa == LOW){
val_aux_piloto=val_aux_piloto - 10;
}
if(val_aux_piloto<val_pot_aceler){
piloto=val_pot_aceler;
}
else{
piloto=val_aux_piloto;
}

porcento = (piloto*100/255);
analogWrite(PWM_motor, piloto);
display_LCD();
}
```

Printed by Books on Demand GmbH, Norderstedt / Germany